SMALL GAS ENGINES

how to repair and maintain them

by Paul Weissler

POPULAR SCIENCE BOOKS

New York

DEDICATION

To Sara,
who loves to help Daddy
type his manuscripts

Library of Congress Catalog Card Number: 75-13334
ISBN: 1-55654-020-5

Eighteenth Printing, 1988

Manufactured in the United States of America

CONTENTS

PREFACE

The gasoline engine has established itself as the portable powerplant of the century. We have learned to produce it in an infinite variety of sizes, and in enormous quantities. It powers our cars, trucks, buses, and motorcycles, as well as a proliferating number of machines.

This book is for the homeowner, so we've devoted our coverage to the three small-gas-engine machines a homeowner is likely to buy. These are the lawn mower, the chain saw, and the snow blower. We have deliberately omitted other machines, yet you'll find the material in this book of great help in repairing and maintaining these others as well.

Among the book's over 400 illustrations, you'll find some manufacturers' drawings. But most of the illustrations are photographs taken specifically for the book. They show you what the parts look like "in the metal," with expert repairmen doing the jobs. The professionals have some special tools. But in most cases screwdrivers, wrenches, pliers, and routine hand tools will be all you'll need. Where special tools are the way to go, the book tells why and where to obtain them.

To work on these machines, you do not have to be a professional mechanic. You don't even have to be an experienced weekend auto mechanic because the typical mower, blower or saw is far less complicated than the engine compartment of your car. There is also a lot less risk of making costly mistakes on these smaller machines. And if you find you cannot make the repair, the replacement component is relatively inexpensive. The final chapter discusses when to repair, when to replace, when to see a professional.

Using this book you will be able to service your mower, blower or saw most of the time. This will save you both money and the inconvenience of leaving the unit in a repair shop. Inasmuch as breakdowns always seem to occur in the middle of the heavy-use season when repair shops are backlogged, shop repair could tie up your equipment for weeks.

Special thanks for their assistance go to the following: Victor Alvarez, Richard Caso, James Combs, Gerald DeRuiter, Thomas Donahue, Martin Ephrain, Lewis Fontan, Dave Kirby, James Morizzo, Vincent Scully, and Harold Vroom—plus Sears, Roebuck and Company, Ronconi Equipment Corporation, Briggs and Stratton, and McCulloch Corporation.

PAUL WEISSLER

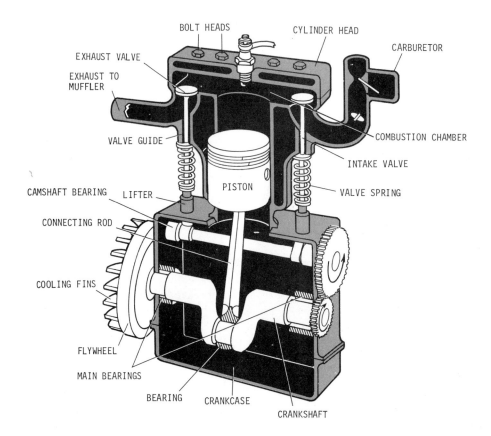

1-1. This cutaway drawing of a one-cylinder four-stroke cycle engine shows how everything works. The turning of the crankshaft moves the piston up and down. Because the crank gear is meshed with the camshaft gear, the camshaft turns, and its cam lobes open each valve at the appropriate time. When the cam lobe rotates beyond the valve lifter, a spring pulls the valve down into the closed position. Notice that each valve controls the flow through passages. One passage leads to the muffler (exhaust) the other to the carburetor (intake).

How Small
Gas Engines Operate

All gas-powered mowers, blowers and saws use a piston engine that is similar in significant respects to those used on automobiles. There are differences, however, most notably in the use of two-cycle engines in chain saws and a few mowers. Let's begin at the beginning and see how the two-cycle and more common four-cycle engines work. This will help you understand what's happening when an engine doesn't run.

The engine develops power by burning a mixture of gasoline and air in a small enclosure called a combustion chamber, as shown in 1–1. As the mixture burns, it becomes very hot and expands, just as mercury in a thermometer expands and pushes its way up the tube when its temperature rises.

The combustion chamber is sealed on three sides, so the expanding gas mixture can push its way in only one direction, downward on a plug—called a piston—which has a close-sliding fit in a cylinder. The downward push on the piston is mechanical energy. When we have circular energy we can turn a lawn mower blade, a chain saw, a snow blower auger, or the wheels of a car.

In the conversion, the piston is attached to a connecting rod, which is in turn attached to a crankshaft with offset sections. A crankshaft functions much like the pedals and main sprocket on a bicycle. When you pedal a bike, the downward pressure of your foot on the pedal is converted into circular movement by the pedal shaft. Your foot pressure is similar to the energy created by the burning fuel mixture. The pedal performs the function of the piston and connecting rod, and the pedal shaft is the equivalent of the crankshaft.

The metal part in which the cylinder is bored is called the engine block, and the lower section in which the crankshaft is mounted is called the crankcase. The combustion chamber above the cylinder is formed in a metal cover for the cylinder, called a cylinder head.

As the piston connecting rod is forced down, and it pushes on the crankshaft, it must pivot back and forth. To permit this movement, the rod is mounted in bearings, one in the piston, the other at its connection point to the crankshaft. There are many types of bearings, but in all cases their function is to support any type of moving part that is under load. In the case of a connecting rod, the load is from the downward moving piston.

A bearing is round and super-smooth, and the part that bears against it also must be smooth. The combination of smooth surfaces is not enough to eliminate friction, so oil must be able to get between the bearing and the part it supports to reduce friction. The most common type of bearing is the plain design, a smooth ring or perhaps two half-shells that form a complete ring, as in 1-1.

Although parts that bolt together are machined carefully for a tight fit, machining alone is not enough. A seal must often be placed between them to prevent leakage of air, fuel or oil. When the seal is a flat piece of material, it is called a gasket. Common gasket materials include synthetic rubber, cork, fiber, asbestos, soft metal and combinations of these. A gasket, for example, is used between the cylinder head and engine block. Appropriately, it's called the cylinder head gasket.

Now let's take a closer look at the gasoline engine's actual operation, which may be either of two types: the two-stroke cycle or the four-stroke.

TWO-STROKE

The term two-stroke cycle means that the engine develops a power impulse every time the piston moves down. The cylinder normally has two ports, or passages, one (called the intake port) to admit the air-fuel mixture, the other to allow burned gases to escape to the atmosphere. These ports are covered and uncovered by the piston as it moves up and down.

When the piston moves upward, the space it occupied in the lower part of the engine block becomes a vacuum. Air rushes in to fill the void, but before it can get in, it must pass through an atomizer called a carburetor, where it picks up fuel droplets. The air pushes open a spring metal flapper over an opening in the crankcase and with the fuel enters the crankcase.

When the piston moves down, it pushes both against the connecting rod and crankshaft, and the air-fuel mixture as well, partly compressing it. At a certain point, the piston uncovers the intake port. This port leads from the crankcase to the cylinder above the piston, permitting the compressed air-fuel mixture in the crankcase to flow into the cylinder.

Now let's look at an actual power cycle in 1–2, beginning with the piston in the lowest portion of its up-and-down stroke in the cylinder. The air-fuel mixture is flowing in and beginning to push burned exhaust gases out the exhaust port, which also is uncovered.

TWO-STROKE CYCLE ENGINE

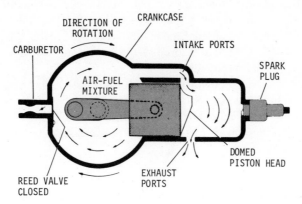

1–2. The two-stroke cycle engine differs from the four-stroke in that it produces power on every downward stroke of the piston. Here ports allow intake of an air-fuel mixture and exhaust of burned gases, as the ports are uncovered and covered by the piston during its strokes. (Ports in the two-cycle perform the function of valves in the four-cycle.) In this design a reed valve covers a passage from the carburetor to the crankcase, but it is the covering and uncovering of ports by the piston that regulates the flow of fuel mixture and exhaust gases in the cylinder and combustion chamber. This drawing shows the piston going down on a power stroke, uncovering the ports. The downward movement increases pressure in the crankcase, forcing the reed valve closed and allowing the air-fuel mixture to flow up into the cylinder. This flow helps push burned exhaust gases out the exhaust port. The dome shape on the piston head helps direct the flow of gases in the cylinder. When the piston rises, it creates a vacuum in the crankcase, which draws open the reed valve to admit an air-fuel mixture from the carburetor.

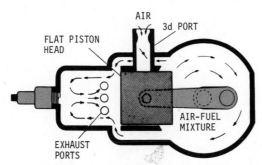

1–3. In this two-stroke design, there is no reed valve, just a third port on the side of the cylinder, through which the air-fuel mixture flows. When the piston rises, creating a vacuum in the crankcase, it also uncovers the third port, permitting the air-fuel mixture to flow into the crankcase. When the piston goes down, as shown in the drawing, it closes the air-fuel mixture port and creates pressure in the crankcase that forces the mixture up into the cylinder. The piston then rises to close off the intake and exhaust ports and compress the air-fuel mixture. Some two-cycle engines have reed valves at this third port to admit a bit more air-fuel mixture for extra power.

The piston begins to move up, simultaneously completing the job of pushing the burned exhaust gases out of the exhaust port, and compressing the air-fuel mixture in the cylinder. When the piston reaches the top of the cylinder, the piston is covering two ports, and the air-fuel mixture is highly compressed. At this point a spark plug, threaded into the combustion chamber, delivers a spark that ignites the mix. The greater the amount of compression, the greater the force of the explosion, and the greater the downward pressure on the piston.

The piston is forced downward and transfers the force through the connecting rod to the crankshaft, turning it. The downward moving piston also uncovers the exhaust port, then the intake port and again begins the job of compressing the air-fuel mixture in the crankcase, to force it to flow into the cylinder above.

Although most two-cycle engines use the flapper valve, called a reed, in the crankcase, some engines do not. They have a third port, covered and uncovered by the piston, that permits the air-fuel mixture to flow into the void in the crankcase created by the upward moving piston. See 1–3.

FOUR-STROKE CYCLE ENGINE

The four-stroke cycle engine develops one power stroke for every four movements of the piston (two up and two down). This type might seem to be a waste of motion as well as parts, for it requires many more parts. However, it has many advantages, particularly in larger engines where compactness is not as significant a factor.

The four-stroke engine does not have a reed, and the air-fuel mixture does not pass through the crankcase. Instead, there are two valves, as in 1–1, one that opens and closes a passage from the carburetor, another that opens and closes a passage to the exhaust system. The valves are operated by the camshaft, a shaft with teardrop-shaped lobes that push the valves open, and at appropriate times, allow springs to close them. The camshaft has a gear at one end, which meshes with a gear on the crankshaft. The gear on the camshaft has twice as many teeth as the crankshaft gear, so that for every complete revolution of the crankshaft, the camshaft turns 180 degrees. This means that each valve opens and closes just once during two revolutions of the crankshaft, which is exactly what's needed for a four-stroke cycle.

The valves in the typical four-stroke lawn mower or snow blower engine are located in the block. This is an antiquated automotive design, but it's good enough for mowers and blowers. There are a few four-strokers with valves in the cylinder head, a popular automotive design, shown in 1–4. In this case the camshaft lobes push on a long rod, called a pushrod, which pivots a see-saw-like part called a rocker arm.

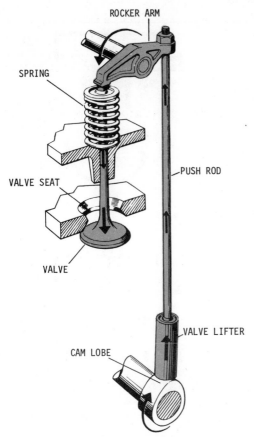

ROCKER ARM

SPRING

VALVE SEAT

VALVE

PUSH ROD

VALVE LIFTER

CAM LOBE

1–4. Here an overhead valve is opened as the cam lobe bears up against the valve lifter and push rod, and therby tips the rocker arm. When the lobe rotates beyond the lifter, the spring pulls the valve up (to closed position). This tips the rocker arm back.

The operation of any four-stroke cycle engine, regardless of valve location, is the same. Let's look at 1–5, with the piston going down in what is called the intake stroke.

The dropping of the piston creates a void in the cylinder, and the camshaft opens the intake valve. Air rushes through the carburetor to fill that void, pulling fuel droplets with it, into the cylinder. When the piston is near the bottom of the cylinder, the camshaft closes the intake valve.

The piston begins rising, and when it reaches the top, it has compressed the air-fuel mixture into the little recess above the piston, the combustion

FOUR-STROKE CYCLE ENGINE:

1–5. This is the first of a series of drawings showing the operation of the four-stroke cycle. It shows the engine during its intake stroke, when the air-fuel mixture is admitted to the combustion chamber. The turning of the crankshaft pulls the piston down, creating a vacuum in the cylinder above the piston. The air-fuel mixture is drawn in to fill this vacuum. During this stroke, the camshaft lobe for the intake valve is in the position that forces the valve up, off its seat, to open and admit the mixture. Here the camshaft lobe for the exhaust valve is closed.

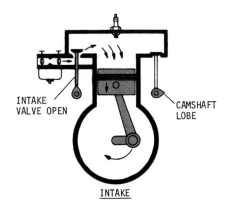

1–6. This is the compression stroke of the four-stroke cycle engine. The crankshaft turns and forces the piston up, compressing the air-fuel mixture. The intake valve closes at this time, and the exhaust valve remains closed. When the piston is at the top of the cylinder, the spark plug ignites the air-fuel mixture, which explodes and expands, forcing the piston down.

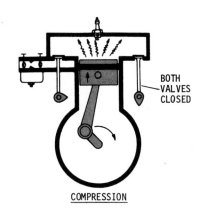

chamber. This upward movement, shown in 1–6, is called the compression stroke.

In 1–7 the spark plug ignites the mixture, which explodes, forcing the piston down in what is called the power stroke.

When the piston rises, as in 1–8, the camshaft opens the exhaust valve and the piston pushes the exhaust into the exhaust system.

FLYWHEEL

To smooth out the movement of the crankshaft and keep it rotating in between power strokes of a two- or four-cycle engine, a heavy flywheel is attached to one end, as shown earlier in 1–1.

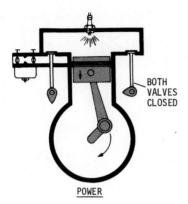

POWER

1–7. The piston is being forced down by the expansion of the exploding air-fuel mixture in what is called the power stroke. Both valves remain closed.

BOTH
VALVES
CLOSED

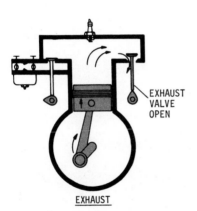

EXHAUST

1–8. The crankshaft continues to turn and the piston rises again, this time on the final stroke of its four-stroke cycle—exhaust. The camshaft lobe for the exhaust valve forces the valve open. The rising of the piston pushes the exhaust out the exhaust valve, through a passage into the muffler, and from the muffler into the atmosphere. The piston will then begin a new four-stroke cycle, as shown in 1–5.

EXHAUST
VALVE
OPEN

The flywheel is an important part of any engine, but it is especially important to the small gas engine. It has a raised hub (of varying designs) in the center, which the starter engages. With manual-start engines, when you pull the starter cord, you are spinning the flywheel. An electric starter, as shown in 1–9, may engage the flywheel hub or spin a flywheel by means of a gear arrangement—one gear on the starter, another on the circumference of the flywheel.

Spinning the flywheel turns the crankshaft, which moves the pistons up and down and, in four-stroke engines, also turns the camshaft to operate the valves. Once the engine fires on its own, you release the starter. An on-the-engine electric starter automatically disengages, forced away by the flywheel, which begins spinning much faster under power from the pistons.

The flywheel is also the heart of the small gas engine's ignition system. Built into the flywheel circumference are several permanent magnets, which provide the magnetic force that the ignition system converts into electrical energy. It is not a purpose of this book to discuss the relationship between magnetism and electricity, but an acquaintance with some aspects helps in understanding how the ignition system gets charged up. Let's begin with a basic explanation of the electric circuit.

AN ELECTRIC CIRCUIT

Without trying to make an electrician out of anyone, let's take a quick run through the basics of an electrical circuit. Unless you know this, such concepts as an electrical ground and short circuit will be very foreign to you, and you may miss something obvious when troubleshooting an electrical problem.

The word circuit comes from circle, and what it means in practical terms is that there must be connections from the source of current to the

1-9. This riding mower has a gear on an electric starter motor that engages a gear on the circumference of the flywheel.

users of the current, then back to the source. Electricity travels in only one direction, so the wire that goes to the source cannot be used as the return.

The simplest circuit is shown in 1–10. Current leaves a terminal on the battery and goes through the wire to the light bulb, a device that restricts the current flow so sharply that the wire inside the bulb becomes hot and glows. When the current passes through the restrictive wire (called a filament in the light bulb), it continues through a second segment of wire back to a second terminal on the battery.

If any part of the circuit is broken, the current flow stops and the bulb will not light. Normally the filament burns out eventually, but the bulb also would not light if either the first or second segment of the wiring between bulb and battery broke. Note that even if the wire from battery to bulb were intact, the bulb would not work if the return wire broke. A break any place in a circuit is called an open circuit; such breaks usually occur in the wiring. Wires normally are covered with insulating material to hold in the electricity, so if the metal strands inside (called the conductor) were to break, you might not see the problem by merely looking at the wire.

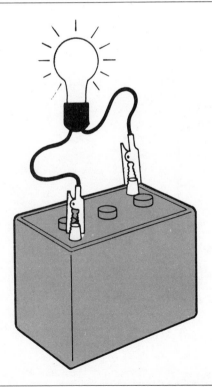

1–10. This is a complete circuit. The wire runs from one battery post through the bulb and to the other battery post.

GROUND CIRCUITS

Metals are the best conductors of electricity, and although copper is the most popular, aluminum is fine and even steel and cast-iron do a good job. This fact permits the elimination of a whole lot of wiring by means of the electrical ground circuit.

The best example is in the automobile. The car battery has two posts, and three cables. One cable goes from a post to the starter motor, whose terminal also serves as a connection point for the other users of current in a car. The other two cables are connected from the other post to the car chassis and to the engine. The chassis and engine are the electrical ground. See 1–11.

GROUNDED CIRCUITS

1–11. This is a ground circuit. One post of the battery is grounded to the chassis and the engine with a two-cable arrangement. The other post has a cable that runs to the starter and to a bulb, completing the circuit without the need for additional cable.

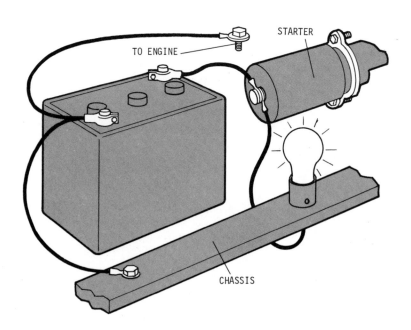

TO ENGINE

STARTER

CHASSIS

The starter motor is bolted to the engine, and the mounting bolts provide a solid metal connection to the ground circuit, so no return wire is needed here. A wire goes to a light bulb, which has a tab that touches a metal part of the socket, which is held to the car body by screws. This metal connection is also an electrical ground.

The engine and chassis of the car, therefore, are one giant return wire for the current.

If part of the insulation came off part of the wiring and bare wire touched engine or chassis, the circuit would be interrupted, because the current would take the path of least resistance. This condition is called a short circuit. See 1–12.

In the typical small gas engine, the uses of the ground circuit are some-

1–12. A short circuit may occur in a ground circuit if the wire to the starter motor, bulb, or other consumer of electricity is chafed and touches the chassis as shown. In this case the current completes the circuit to ground without going through the starter or the bulb.

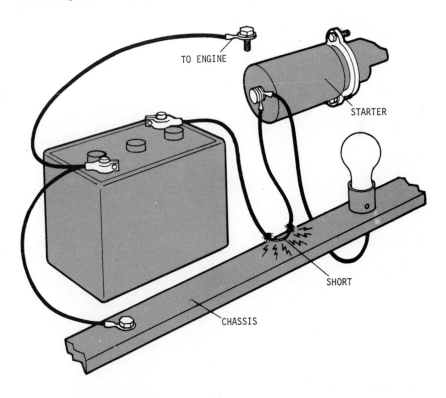

TO ENGINE

STARTER

SHORT

CHASSIS

what less numerous, but they are there. For example, the magneto, a key part of the ignition system, is grounded to the engine block by its mounting screws and so is one of the ignition breaker points. When the points are closed, the circuit between magneto and breaker points is completed because both parts are grounded to the engine.

When the spark plug current jumps the gap in the plug, it actually is crossing an air gap to an electrical connector that is part of the metal shell of the spark plug, and that shell is threaded into the cylinder head, which is bolted to the engine block. These two metal to metal connections complete the ignition system's high voltage ground circuit.

IGNITION SYSTEM

A close look at how the ignition system works will demonstrate the importance of the flywheel magnets. They are used in the conventional magneto ignition system and in the typical new transistorized ignition systems used on a few premium mowers and blowers.

The conventional system features a coil, called a magneto, mounted on the engine very close to the flywheel. Like an automotive ignition coil, it

IGNITION SYSTEM OPERATION

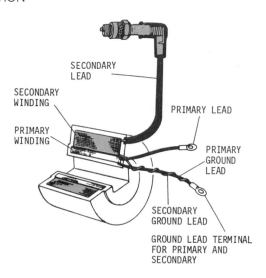

1-13. This is the magneto coil, which consists of a thick-wire primary winding and a thin-wire secondary winding. The primary circuit is grounded at one end by the primary ground lead, which is screwed onto the engine block. The other end goes to the breaker points, and the circuit is complete when the breaker points are closed. The secondary circuit is grounded by the secondary ground lead (usually the same terminal as the primary) at one end and goes to the spark plug at the other end. When the current in the secondary jumps across the spark plug electrodes, it again reaches ground, completing a circuit.

SECONDARY LEAD

SECONDARY WINDING

PRIMARY WINDING

PRIMARY LEAD

PRIMARY GROUND LEAD

SECONDARY GROUND LEAD

GROUND LEAD TERMINAL FOR PRIMARY AND SECONDARY

has two windings, one of relatively thick wire, another of many more turns of thinner wire. See 1–13.

The thicker wire is connected to a set of breaker points and a condenser, just as in older automobiles, as shown in 1–14 and 14a. The breaker points are simply two electrical contacts—a switch. One is fixed, the other is movable. A lobe, on the crankshaft, somewhat similar to those on the camshaft that operate the valves, pushes open the points once every crankshaft revolution. When the lobe spins away from the movable point, a spring pulls it back into contact with the fixed point.

The points are wired into an electrical shock absorber called a condenser, which absorbs stray high voltage during the firing of the spark plug, to prevent premature burning of the points.

When the points are closed, they complete a circuit to electrical ground. When they are opened by the crankshaft lobe, they interrupt the circuit. Here's how they are used to create high-voltage electricity that ignites the air-fuel mixture in the cylinders: As the flywheel spins, in 1–14 and 14a, the magnets pass the coil, which converts the magnetism to electrical energy. The electricity flows through the thick wiring of the coil to the ignition breaker points, through them (when they are closed) to electrical ground, completing what is called the primary circuit.

When the breaker points open, as in 1–15 and 15a, the interruption in the circuit causes the primary circuit to collapse. The electricity does not disappear, but is transferred to the thin-wire winding. Because the second winding is thinner, the same amount of current that flowed under low pressure through the thick winding must now be compressed in the thin winding. Electrical pressure is called voltage, and what happens is that the voltage increases tremendously in rough proportion to the number of wire windings in the primary circuit versus those in the secondary. A typical magneto coil may have 150 turns of primary winding and 10,000 turns of secondary, ratio of 1 to 70. If primary voltage were 300–400 volts, the secondary might be as high as 20,000.

The principle of forcing a low voltage circuit to collapse on one of many more windings of wire to produce high voltage is basically what is applied in electrical utility transformers. The reason that the current transfers from the thick wire to the thin wiring is that electricity takes the path of least resistance to find a way to complete a circuit.

The ignition system is designed so that this electricity will complete a circuit through the spark plug. When the high voltage forms in the secondary, it looks for the easy way out. Some of it tries to jump across the breaker points, but the condenser temporarily absorbs it. The rest travels along a wire to the spark plug, which has two tips that are positioned in the combustion chamber.

These tips are separated from each other, but one is part of the plug's *metal* shell. High-voltage electricity travels down to the core tip, then

MAGNETO IGNITION SYSTEM (POINTS CLOSED)

1–14 and 14a. Here's the magneto ignition system in action. As the flywheel magnet passes by the coil, the magnet in motion creates an electric current in the primary circuit, which is completed to ground at both ends because the crankshaft lobe has allowed the movable breaker point to close.

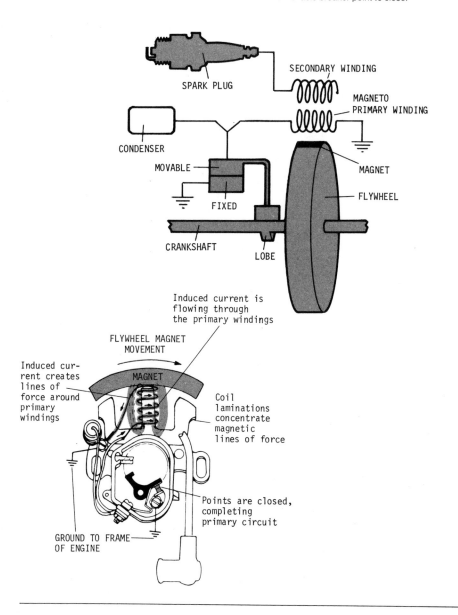

MAGNETO IGNITION SYSTEM (POINTS OPEN)

1–15 and 15a. Here the magnet has passed by. The crankshaft lobe opens the breaker points, causing the electromagnetically charged field of the primary circuit to collapse upon the secondary circuit. This results in high-voltage current to the spark plug—the terminal of the secondary circuit.

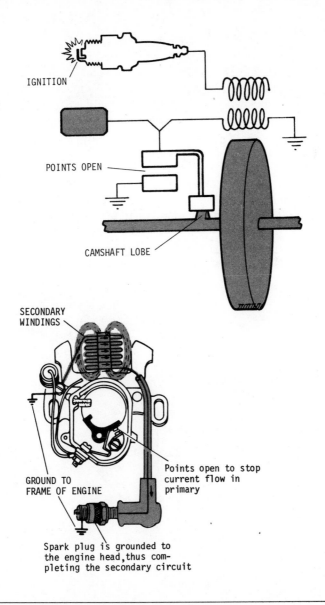

IGNITION

POINTS OPEN

CAMSHAFT LOBE

SECONDARY WINDINGS

GROUND TO FRAME OF ENGINE

Points open to stop current flow in primary

Spark plug is grounded to the engine head, thus completing the secondary circuit

jumps a small gap (perhaps .025 inch) to the other tip. The jumping of the electricity from one tip to the other, through an air gap, is what we call a spark, and it ignites the air-fuel mixture.

When the current reaches the second tip, it has completed its journey, because the plug is threaded into the cylinder head and electrical ground.

TRANSISTORIZED IGNITION

All new cars have ignition systems in which the breaker points and condenser are eliminated by transistorized circuitry, and if you want, you can buy some mowers and blowers with this feature. It's really an unnecessary expense for the homeowner, for the additional cost of transistorized circuits cannot be justified on equipment used so relatively little. It can barely be cost-effective in an automobile, and the major reason for its use in cars is to maintain low exhaust emissions.

The transistorized design used in mowers and blowers still is a form of magneto ignition. A typical design, shown in 1–16, is by Tecumseh. Here the permanent magnets remain in the flywheel, and a trigger coil and input coil are positioned just a few thousandths of an inch away on the en-

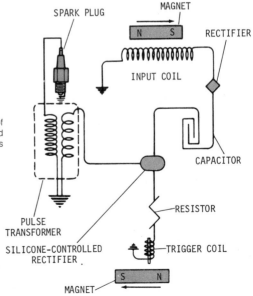

1–16. This is the schematic layout of the transistorized ignition system used on Tecumseh engines. Operation is explained on accompanying pages.

gine block. The conventional coil with its two windings and its wire to the spark plug is renamed a pulse transformer and is spliced into a circuit between the input and trigger coils.

As the flywheel spins, it passes the input coil, a thick-wire winding that converts the magnetism into a low-voltage alternating current. The current passes through a rectifier, a one-way solid-state device that converts the AC to DC, which then passes into a condenser for storage. As the illustration shows, there is a rectifier on each wire to the condenser. The second rectifier can be open or closed, but at this time it's closed, preventing current from getting out of the condenser. In electronic circuitry, the condenser is called a capacitor, and the system is therefore called a capacitive-discharge design, for the spark will occur at the plug shortly after the capacitor discharges the current it is storing. But we're a bit ahead of ourselves.

After the flywheel passes the input coil to charge the capacitor, it reaches the trigger coil, in which its magnetism induces a small amount of current which closes another rectifier that serves as a solid-state switch. When this rectifier closes, the capacitor can discharge, and discharge it does into the pulse transformer's thick-wire winding. The flywheel magnet is now past the trigger coil, so current to the second rectifier is shut off, making it automatically pop open. This is equivalent to the opening of breaker points, and the circuit in the thick-wire winding collapses on the pulse transformer's thin-wire winding, resulting in stepped up voltage, which is discharged to the spark plug.

FUEL SYSTEM

An engine really runs primarily on air, about 14 parts of air to one of gasoline. The job of the fuel system, therefore, is to first mix air and fuel in proper proportions and then deliver it to the combustion chamber.

The carburetor is the key component. It mixes the fuel and air, and in some small engines, it also houses the fuel pump, which draws fuel from the tank and delivers it to the carburetor.

The typical small engine carburetor is of simple design, simple that is, if you're used to automotive carburetors. If you were able to wade your way through engine and ignition system operation, however, you can understand carburetion too.

Begin by thinking of a perfume atomizer. You squeeze the bulb and a spray of perfume comes out. If the bowl contained gasoline, you'd get a spray mixture of air and gasoline droplets. The atomizer looks simple but you probably never thought about how it works, so as a fringe benefit of learning about small gas engines, you can also understand this boudoir essential.

With the atomizer, squeezing the bulb forces air through a horizontal tube, shown in 1–17. This creates a low pressure zone over a jet of a connecting tube that extends down into the perfume. Since the air in the atomizer bottle itself is at normal air pressure (14.7 pounds per square inch at sea level, a bit less at higher altitudes), it forces the perfume up the tube toward the lower pressure. Then the air stream picks up the droplets and expels them as spray.

This is really what a carburetor is all about. But instead of perfume, its jet carries gasoline. Instead of blowing air past the tip of the jet by means of a bulb, the carburetor has a specially-shaped cylinder called an air horn through which the engine applies vacuum, as in 1–18.

The two-cycle engine uses vacuum created in the crankcase when the piston rises. That vacuum pulls open the reed valve and draws in air from the carburetor air horn to create a low pressure area there. As outside air rushes in to fill the vacuum, it creates a special little low pressure zone around the tip of the jet, drawing fuel out in the form of droplets that it carries into the crankcase.

The four-cycle engine uses vacuum created in the cylinder when the piston goes down. Instead of flowing into the crankcase, the air-fuel mixture goes directly into the cylinder when the intake valve opens. Aside from these differences, the method of supplying fuel to these two engines is essentially the same.

The air flow through the carburetor determines the amount of air-fuel mixture the engine will receive. To control that flow, there is a circular plate called the throttle, which is hinged in the center of the air horn. When you operate the throttle control (or step on the gas pedal in a car) you pivot the circular plate to the vertical position to permit maximum air-fuel mixture flow.

It's also important to understand how the fuel gets to the carburetor and how it is metered into the jet. For the little mechanisms that do these jobs are the key moving parts in the carburetor and are subject to failure. These parts must function properly, or else either of two problems will occur: 1) Too little fuel will get into the cylinder, and the engine will starve and stall. 2) Or too much fuel will get in, causing the engine to flood and then stall. (The right amount for an explosive mixture is in a narrow range.)

The fuel tank houses the gasoline. And in the simplest setups it is mounted above the carburetor and connected to it by a tube. Fuel flows by gravity from tank to carburetor, which has a small bowl to store enough to keep the engine supplied for perhaps a minute. This system works fine for household-type mowers and blowers.

Another basic design, perhaps the simplest, is the suction lift carburetor, shown in 1–19. This carburetor consists of a jet, an adjustable tapered needle that threads into it (to adjust fuel flow), a throttle, a choke, an air horn, and one or two suction pipes ("fuel drinking straws") that project

CARBURETOR PRINCIPLES

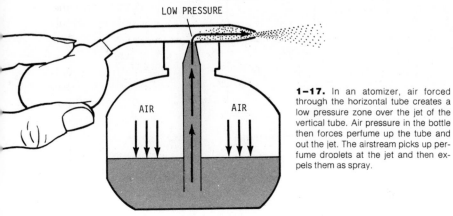

1—17. In an atomizer, air forced through the horizontal tube creates a low pressure zone over the jet of the vertical tube. Air pressure in the bottle then forces perfume up the tube and out the jet. The airstream picks up perfume droplets at the jet and then expels them as spray.

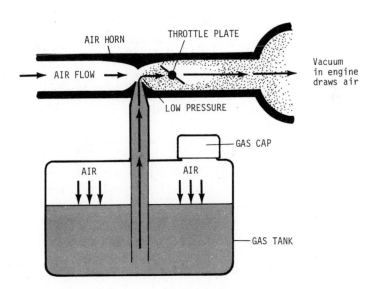

1—18. In a carburetor, the airstream results from engine vacuum whose force is affected by the position of the throttle plate. Air pressure in the gas tank forces gas into an air horn where the air and gas mix and rush to the combustion chamber.

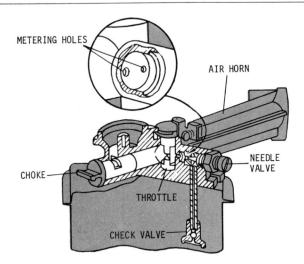

METERING HOLES

AIR HORN

NEEDLE VALVE

CHOKE

THROTTLE

CHECK VALVE

1–19. Probably the most basic small gas engine carburetor is the one shown in this drawing. Fuel is drawn up the pipe through the check valve into the carburetor, where its flow rate is controlled by the tapered needle's position. The fuel then is drawn into the air horn, where it mixes with the air and flows into the engine. If the choke is closed, the air horn is restricted, and less air can flow through. Since the fuel flow is little changed, the mixture therefore is richer with fuel and allows easier starting.

down into the gas tank. The vacuum in the carburetor air horn sucks fuel up the straw through the jet into the air horn.

In many mowers and blowers, however, gravity feed isn't possible because the gas tank can't be mounted high enough, and the simple suction lift doesn't provide the fuel control to enable the engine to function well at all speeds. In these cases more complex fuel pumping and metering systems are used. These are both built into the carburetors on the small engines you are likely to have on your mower or blower.

In the chain saw, clearly, the varied working angles make a gravity feed system impractical. And to provide good fuel supply under all conditions, the simple suction lift wouldn't be much good either.

The on-carburetor pump is a piece of flexible plastic into which are cut two C-shaped flaps that move up and down in response to pulses of vacuum in the engine. They cover and uncover passages from the fuel tank and to the carburetor's fuel delivery system, where fuel is metered into the air horn.

In some carburetors, the crankcase pressure and vacuum simply move a one-piece diaphragm, which draws open and forces closed inlet and outlet ball-type valve. This design consists of a steel ball in a specially-shaped fitting threaded into the passage. When the ball is moved one way, it seals the passage; when it is moved the other way, fuel can flow past it.

Once the fuel is in the carburetor, either of two methods is used to control the storage and metering.

On most mowers and blowers, a float system is used, much like the one used in a toilet tank. As shown in 1–20, a hinged float with a projecting arm drops when the fuel level in the carburetor bowl is low, permitting a tapered needle to come off its seat, opening a passage to the bowl. The fuel flows in, causing the float to rise. When the float reaches a designated level, it pushes the needle back into its seat, shutting off the fuel flow. The float insures an adequate supply and the jet draws from the float bowl as necessary.

On chain saws the float system won't work, because the chain saw is used at so many different angles that the float wouldn't keep the bowl properly filled at all times.

Instead, there are floatless designs in use, featuring a diaphragm that moves a tapered needle valve. When the crankcase creates a vacuum, it draws the carburetor diaphragm; this creates a vacuum that also draws the needle off its seat, permitting fuel to flow through a jet into the air horn, to mix with the inrushing air. As shown in 1–21, diaphragms may work in many ways. Also see 1–22 through 1–25.

OTHER CARBURETOR PARTS

The typical small engine carburetor also has a choke, which may be a round plate hinged in the air horn, or a disc that can be pivoted to cover

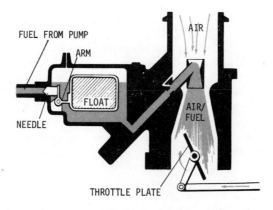

1–20. As fuel is drawn from the carburetor bowl and out the jet, the float drops, causing its arm to unseat the needle. This allows the fuel to replenish the bowl. As the fuel level then rises, the float arm gradually reseats the needle, shutting off supply.

DIAPHRAGM APPLICATIONS

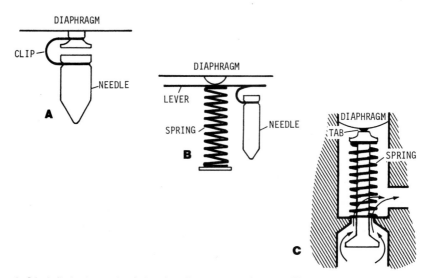

1-21. A diaphragm can be designed to affect a variety of linkups. **(A)** it may be connected to the needle itself. **(B)** it may merely touch a lever, keeping the needle closed. When the diaphragm is drawn away by engine vacuum, the lever rises, permitting fuel to flow past the needle. **(C)** The diaphragm may bear against a spring-loaded needle valve. Engine pressure pushes the diaphragm down, and a contact tab in the center of the diaphragm pushes the needle down against spring pressure, opening it to permit fuel flow. Diaphragm mechanisls, it should be noted, are not limited to chain saws. They are coming into wider use on mower and blower engines.

WHAT IS A DIAPHRAGM?

A diaphragm is the operational heart of many mechanical devices. It is a flat sheet that flexes in response to vacuum or pressure. If pressure is exerted from above, it flexes downward. If vacuum is applied above, it flexes upward. When the diaphragm is mounted so that the area below it is sealed, the diaphragm creates the same pressure below as is exerted above. Example: If the diaphragm is drawn upward by a vacuum, the area below it becomes larger, but contains the same amount of air. Therefore, it too becomes a partial vacuum because the air is thinner (therefore weighs less) than the air outside. If the diaphragm is pushed downward by pressure, the area below becomes smaller, but because the quantity of air remains the same, the air is under some pressure too. In a small engine carburetor the diaphragm therefore becomes a useful device that collaborates with valves to move fuel. It also is the heart of the automobile fuel pump.

the air horn at the top. In either case, closing the choke plate restricts the air flow through the air horn, so that the air-fuel mixture is exceptionally rich with fuel. This is necessary for cold starting on small gas engines, as it is on large passenger car powerplants. The reason for the choke involves the components of gasoline and the metal temperatures of the fuel system and engine. For best combustion the fuel must be capable of becoming a well-mixed vapor very quickly. In a warm engine, the heat emitted by the engine can vaporize the fuel very easily. In a cold engine, however, vaporization is a problem. Gasoline actually is a blend of many types of fuel, some of which vaporize easily, others of which do not. When the mixture is made very rich with fuel, there will be enough of these easily-vaporized components to form a mixture that will burn and produce adequate power when the engine is cold.

The carburetor also may have three fuel jets projecting into the air horn, one for low-speed operation, another for medium-speed (rare) and a third for high-speed.

To insure that the air entering the carburetor and engine is clean (since abrasives in the air cause engine and carburetor wear), an air filter is mounted on top of the carburetor air horn. This filter must be clean. If clogged with dirt particles, it restricts the flow of air (and fuel) into the carburetor, thereby reducing the maximum possible performance.

Fuel system operation is somewhat tiresome to many. This concludes the theoretical section in this chapter. Those who wish some more detail, with some specific examples, will find it in Chapter 7.

GOVERNORS

The purpose of a governor is to keep the engine from destroying itself by running too fast. There is one hard-and-fast rule about them for homeowners: Leave them alone. The only exception is the governor spring on some designs. If you accidentally elongate the spring while removing the carburetor, replace it.

Despite these restrictions, perhaps you would like to know how it works.

The air vane governor, shown in 1–26 and 26a, is the most common one. Its operation is quite simple: An air vane connected to the throttle shaft is positioned near the flywheel whose fins blow air and thus serve as the engine's cooling fan. The faster the flywheel spins, the greater the air flow, which pushes on the vane. At very high flywheel speed, the air flow is sufficient to move the vane. The movement of the vane tends to pull the throttle shaft toward the closed position, stretching the spring which connects the lower part of the throttle shaft and a bracket. As the throttle is pushed slightly closed by the action of the vane, engine speed drops. The

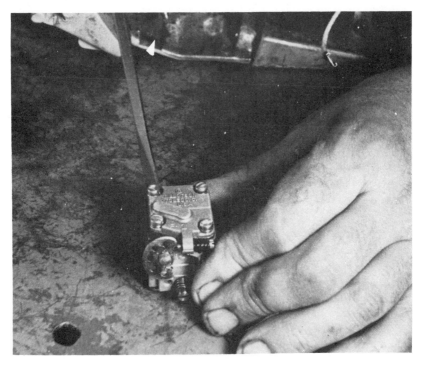

1-22. This is the start of disassembly of a Tillotson diaphragm carburetor for a look inside. The fuel pump section cover is being removed.

1-23. Now the fuel pump section cover of the carburetor is off, showing a gasket at lower center and a diaphragm with C-shaped cutouts to its right.

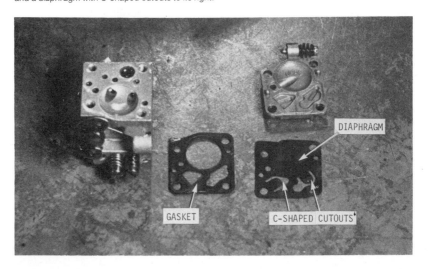

1–24. The fuel metering section cover is being removed. The hole in the center of the cover permits atmospheric pressure to bear on the diaphragm underneath.

1–25. Here the fuel metering section cover is off, showing a gasket with metal cover (upper left) and diaphragm to its right. The top of main body contains a spring-loaded lever and the needle it controls to regulate fuel flow.

AIR VANE GOVERNOR

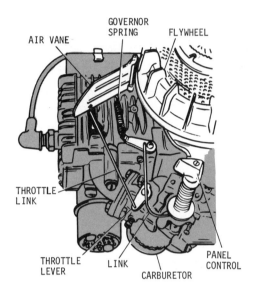

AIR VANE

GOVERNOR SPRING

FLYWHEEL

THROTTLE LINK

THROTTLE LEVER

LINK

CARBURETOR

PANEL CONTROL

1–26 and 26a. When engine speed increases, the flywheel blows more air against the vane, forcing the vane to pivot in a direction that tends to push the throttle closed, slowing the engine. When the engine slows, the air flow against the vane is reduced, and the vane moves back into its normal position. The vane and its calibrated spring will force the throttle toward the closed position only when engine speed has risen above the maximum at which the engine should operate.

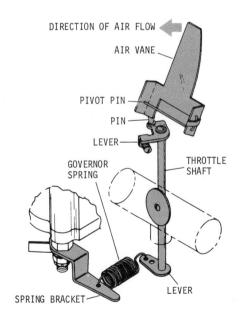

DIRECTION OF AIR FLOW

AIR VANE

PIVOT PIN

PIN

LEVER

THROTTLE SHAFT

GOVERNOR SPRING

LEVER

SPRING BRACKET

spring coils pull together and the vane assumes its normal position. The position of the spring on its bracket is adjustable if necessary, but leave it alone. The spring itself is a carefully calibrated part, so handle with care.

Many engines employ a mechanical governor. This is a centrifugal device mounted on the crankshaft or its own little shaft. There are literally dozens of types of mechanical governors, but a typical one might be a component with counter weights that swing out at high speed and push a link, connected to the throttle linkage, toward the closed position.

The mechanical governor is usually an internal component and doesn't get in the way of routine service. The only exception of note is the governor on the Lawn-Boy mower, which is under the flywheel and must be lifted off the crankshaft before you have access to the points.

How Engines
Accomplish Work

Although there are literally hundreds of manufacturers of lawn mowers, snow blowers and chain saws, the sources of power are relatively few. Briggs and Stratton, for example, makes perhaps two-thirds of the four-stroke cycle engines used in lawn mowers and snow blowers. It should be clear that once you've got a powerplant, almost anyone can make the frame and attachments that enable it to do useful work.

The rotary lawn mower, far and away the most popular type in household use, is a perfect example. Just mount the engine on the housing so that the crankshaft is in a vertical position, bolt a blade onto the bottom end, run the engine and the blade will cut the grass. See 2–1.

Well, it isn't quite that simple, particularly with reel mowers, snow blowers and chain saws, but reasonably close. The engine is the heart of the appliance. You still, however, should know something about the components that convert the spinning of the crankshaft into snow removal and into grass and wood cutting.

One part common to all the appliances but the simple rotary mower is the clutch. As in the automobile, the clutch serves to disconnect the engine from the actual drive mechanisms. In the car, the clutch permits you to stop the car without turning off the engine. In small gas engine equipment, the clutch enables you to start the engine and keep it running without spinning the work-performing parts.

Some small gas engine appliances made for household use, particularly chain saws, have what is called a centrifugal clutch. This type of clutch engages only when engine speed is high enough to create enough centrifugal force in the clutch assembly.

You've seen centrifugal force at work many times. When you stir a cup of coffee, the swirling liquid is impelled away from the center toward the inner circumference of the cup. This shows the effects of centrifugal force.

2-1. Remove the blade of a simple rotary mower and you see the crankshaft of the small gas engine that spins it. The application of power is very direct.

2-2. The chain saw, like the simple rotary mower, applies engine power directly to the working part, in this case a cutting chain. When engine speed is high enough, the friction shoes of the clutch are pushed outward by centrifugal force, and they engage the drum around which the chain is wrapped. The clutch thereby transfers engine power to the chain, which begins to move around its guide bar.

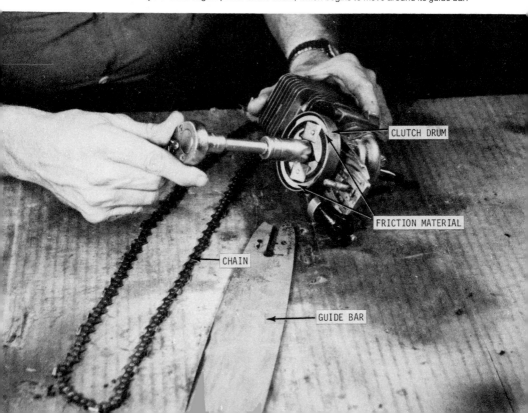

CLUTCH DRUM

FRICTION MATERIAL

CHAIN

GUIDE BAR

The centrifugal clutches shown in 2–2 and 2–3 consist of pieces of friction material mounted on shoes which are held by springs to the crankshaft. Over this assembly is fitted a drum which has a hub with a bearing that rides freely on the crankshaft. The friction shoes are connected to the crankcase.

When the engine idles, the shoes turn with the crankshaft, but the drum remains stationary. As the engine speed increases, centrifugal force pushes the friction shoes against the inner circumference of the drum, thus locking the drum and the shoes. This is automatic clutch action.

The drum has either a projecting shaft or gear-shaped sprocket to which the working parts of the appliance are attached. There may be a sprocket and a chain which are connected to yet another part of the appliance. This permits the engine power to be transferred to a location where it's needed. Another device for power transfer is the drive belt, wrapped around one pulley built onto the clutch drum and a second at the power-using location.

Many mowers and blowers have a manual clutch that you engage by

CENTRIFUGAL CLUTCH

2–3. When the engine idles, the friction shoes rotate with the relatively slow-turning crankshaft. But when speed increases, centrifugal force pushes the shoes against the inner circumference of the drum, thus making shoes and drum lock as one revolving unit. When engine speed decreases, the springs overcome the lessened centrifugal force and disengage the shoes from the drum.

CLUTCH SPIDER AND SPRING ASSEMBLY

CLUTCH AT REST   CLUTCH ENGAGED

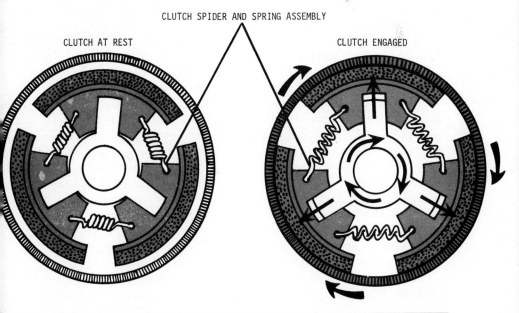

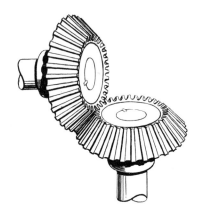

2–4. Bevel gears like these can change the direction of the axis of rotary motion.

moving the handle or pulling a lever. The manual clutch is typical on reel and self-propelled mowers and blowers.

In the reel mower, chain saw and snow blower, the engine is mounted so that the crankshaft is horizontal. At one side is the flywheel; at the other is the clutch with sprocket or pulley.

The saw chain does not merely transfer power; it's also the part that does the job of cutting wood. It fits onto the clutch sprocket at the crankshaft and around an elliptical guide bar, which has a recess or groove, in which the chain rides. When you turn a screw to adjust tension on the chain saw chain, all you are doing is moving the guide bar farther away from the clutch sprocket.

The chain or rubber belt is a power transfer device in the snow blower, reel mower and any type of self-propelled mower, including rotary.

If the crankshaft is horizontal, power-take-off is often very simple: A chain or belt either back to the rear wheels or forward to the front wheels, or to the working parts of the reel mower or snow blower. If the crankshaft is vertical, a horizontal power-take-off shaft may be built into the engine, driven by a bevel gear from the crank. Bevel gears, like those in 2–4, can convert vertical rotary motion into horizontal. Or the crankshaft may be extended, and a gear arrangement attached to provide horizontal rotary motion. In either conversion, a belt or chain carries power to the front or rear wheels.

The methods of engaging and disengaging the self-propulsion system on mowers and blowers also vary, although the control usually is in the handle or in a lever on the handle. One belt system has an additional pulley called the idler. The lever that engages the self-propulsion simply actuates a cable or linkage that pushes the idler pulley in such a way as to apply tension to the belt. Releasing the self-propulsion lever allows the idler pulley to spring back, slackening the drive belt so it is too loose to transmit power.

Chain drive systems on mowers and blowers normally use a type of clutch that locks a sprocket to one end of its shaft. This sprocket is mounted on a bearing so that when the clutch is released, the sprocket can turn freely.

The variety of mechanical designs, power-take-off and self-propulsion is so great that a book could be written on that subject alone. A close look at your machine while the system is being engaged will show you what type you have. Most of them are anything but sophisticated.

Although the chain saw and rotary mower apply their power rather directly, the snow blower and reel mower transmit it to working parts at the front. See 2–5 and 2–6.

In the reel mower, the power is transmitted to a shaft with S-shaped blades mounted on it. As the blades turn they pass a horizontally-mounted cutting bar. The grass is impinged between reel and bar and neatly cut, in an action very similar to that performed by a pair of scissors.

The reel mower costs much more than a rotary, and often comes with a seemingly low-performance engine, but appearances are deceptive. The rotary needs a lot more power to spin the blade very fast, for its cutting action is more like a karate chop. The actual cutting is done only at each tip, so the cutting surface is far less than that of the reel type.

The snow blower engine transmits its power to a shaft with a large auger, which looks like a drill bit with sections cut out. The spiral shape enables it to dig into the snow, scoop it up and spin it out through the discharge chute.

THE ENGINE LAYOUT

Now that you know how the engine is used to power the appliance, you can take a close look at your equipment and the layout of the engine. A general familiarity with the positioning of components takes much of the complexity out of the unit itself. You learn to look at individual sections and components, rather than the entire unit.

In certain respects, most brands of chain saws are laid out the same way. Viewed from the rear, the guide bar is at the right front, so that the chain also must be on the right side. This means that the clutch and sprocket also must be located there.

The flywheel starter and ignition, therefore, are normally on the left side. There are exceptions, such as McCulloch's with the starter on the clutch, or a gear-driven saw with the starter on the transmission. However, the most popular homeowner's models, including McCulloch's, are in the

2–5. Connected to a small gas engine, the chain on the right spins this huge auger, which hurls snow up through the chute.

2–6. Like the auger on the snow blower, the reel on this mower is chain driven. The drive is supplied by a small gas engine in the rear.

first category. Although the piston and connecting rod could be at the center front or center rear, from a standpoint of operator handling, there's only one logical location, mechanically—toward the rear. This places the spark plug at the rear end.

All popular chain saws have a two-stroke cycle engine, because it's smaller and lighter, and all have a diaphragm carburetor with a diaphragm pump built in, to permit using the saw at all angles.

All rotary lawn mowers have a vertical crankshaft engine with a clutch at the lower end of the shaft, to which the blade is mounted by means of an adapter. The flywheel, starter and ignition therefore are at the top. If the fuel tank is at the top of the engine, a gravity feed fuel system is used. If it's on the side, and you find the carburetor mounted on it, a suction lift carburetor is the design. If the tank and carburetor are separated, the carburetor has a built-in fuel pump section and probably a float chamber.

Self-propelled models may have an additional horizontal shaft coming out of the engine, and a sprocket and chain or pulley and belt setup to drive front or rear wheels.

Reel mowers typically have a horizontal crankshaft engine, with the flywheel normally on the right side (viewed from the rear) and the clutch and sprocket on the left. The carburetion arrangement for the rotary also applies to the reel.

The snow blower layout is very similar to that of the reel type mower.

Although the mechanically driven mechanisms of mowers, blowers and chain saws vary greatly and may require detailed inspection just to figure out, the heart of any of these appliances is the small gas engine. It is the component that requires the most routine maintenance and the most repair. This is not to imply you can just forget about everything but the engine. You'll find much on the maintenance of driven systems in Chapter 8.

3

Keeping Your
Jobs Organized

It's impossible to learn once and for all how to repair every small gas engine from any book. What you can learn are the basic operating principles, the general layout of the equipment and the how-to of the most popular equipment currently in use.

Then if a model comes up with some feature you haven't seen before, you merely have to add to your existing knowledge. Perhaps you can figure out what the new widget is, but if you can't, it doesn't keep you from servicing those items you do know. Even if you must remove the unknown part, common sense precautions will keep you from incorrect reinstallation.

There are two possible sources of information: 1) The factory distributor is listed in the yellow pages of the telephone book. He's the man from whom you'll be buying your parts; he's also the one who can keep you up to date. If he can't answer your questions, he can check with a mechanic in his service shop. Unlike many car dealers, the distributors of small gas engine appliances are somewhat free with adjustment specifications and advice. 2) A factory shop manual may be available from the distributor. But the availability of a manual varies from one distributor to another, and from one make to another. Briggs and Stratton manuals are reasonably available almost everywhere, but others are strictly catch-as-catch-can. A caution about shop manuals: *They do not teach you the basics of repair.*

The small gas engine and its accessories are tiny by automotive standards. You can lift the whole appliance if necessary and service anything on a small table. Try doing that with even a Volkswagen four-cylinder engine.

Because the components are small doesn't mean you should lose respect for them. In fact, they deserve extra care because their size makes them easy to lose. Damage to a small part may not be visible. And the correct

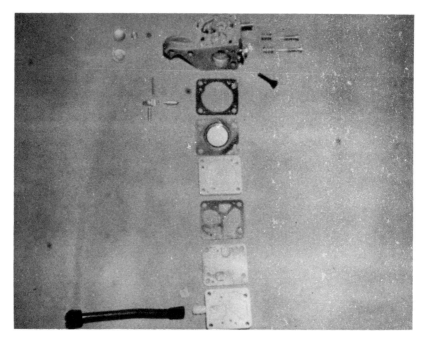

3-1. This little chain saw carburetor has many parts. So unless you make a careful sketch and keep track of all of them, reassembly could be very difficult.

reassembly may not be obvious if you have taken apart a component somewhat casually. See 3-1.

Knowing how to take apart a mechanical component is not the inherent talent of the "handy" person. It is plain old-fashioned experience. If you'd never seen a light bulb before, you'd never know it was threaded into the socket. Nor would you know which way to turn it to loosen or how tight to make the replacement. A mechanic who takes apart a Briggs and Stratton engine or a Tillotson carburetor several times a week soon learns to do things automatically. Unless you have recurring problems with your small gas engine equipment, you'll never reach this stage of expertise. Therefore you must work on the components with care.

Always leave enough time to do the job from start to finish. This is extremely important if you are to be able to remember what goes where for reassembly. Clearly, if you haven't done a job before you really don't know exactly how long it should take, so a good rule is to start a totally unfamiliar job very early in the morning when you have the entire day available, even if it appears to be a 15-minute proposition.

Make sure you have the parts and tools you need right at hand and can

3–2. You can make this engine repair stand yourself from hardwood. The stand helps you work at the optimum angles and helps prevent breakage of parts. This stand was designed by Tecumseh-Lauson engineers, but it will work with other makes of small gas engines.

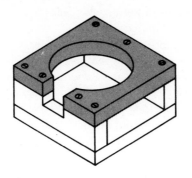

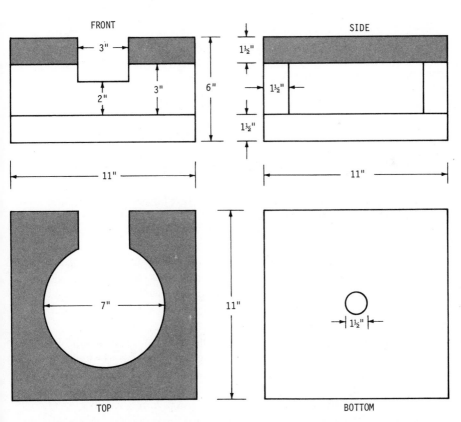

FRONT

SIDE

TOP

BOTTOM

get something you didn't expect to use in a hurry. Except for such minor things as spark plug replacement, you could easily find yourself short a part or a tool. The odds are that the parts will not be available on Sundays, and even Saturdays in some areas. The best procedure is to tell the parts counterman what you're planning to do and have him advise you. He should know if there's some gasket or whatever you're likely to need and whether or not special tools are required.

Buy your parts from an authorized distributor, at the beginning anyway. The parts counterman at factory distributorships are more knowledgeable about the makes they specialize in, and therefore can offer the best advice. They also can order special tools if necessary.

Set up a clean, clear area in which to work. You're less likely to lose parts this way. A discarded soup pot or small pail makes a suitable cleaning pan, and ordinary automotive solvent, such as Gumout, is fine for cleaning small engine parts. You can, if you wish, make an engine holding stand from scrap wood. See 3–2.

Keep all nuts, bolts and other hardware for a single component together. A good procedure is to have a bunch of small envelopes handy, place the parts in each envelope and label the envelope. For examples: *side cover screws, chain oiler, cylinder head bolts.* Or you can wrap them together with masking tape. Another method of holding on to parts, particularly bolts, is to refasten them where possible, after removing a part they held. Example: If you remove the cylinder head held by studs and nuts, rethread the nuts onto the studs immediately after removing the head. Still another method is to push the parts into a thin piece of cardboard as you remove them. This is a very effective technique when you want to keep parts in a very specific order. Example: The side cover of a small gas appliance has different length bolts. To avoid confusion during reassembly, push them into the cardboard in the general pattern of the bolt holes in the cover, and mark one side of the cardboard "top."

Make careful sketches of assemblies as you take them apart, so that you have an accurate reference during reassembly. You don't have to be an artist, but you must know how to look at a part, and you must disassemble slowly. This will pay off in fast reassembly. Some of the things you should look for when you remove a part—and include them in your sketch—are the following:

1. Is one side different from the other? A little washer may actually be concave. A little rubber sealing ring may have a lip on one side only. A spring may be coiled differently at one end than at the other. One bracket may have a different size hole at one end than the other. The manufacturer had a reason for the variation.

2. Are there adjacent holes into which a part could erroneously be fitted? Or are there similar parts that could be mixed up? A simple example

of this is the part of the speed control linkage into which the throttle and governor springs are fitted. The part may have three or four holes, only one or two of which are used. You must hook the correct hole, or the engine will not perform properly. So make sure your sketch provides the information you'll need.

3. Does more than one part fit into a single hole? A chain saw oiler, for example, may have five parts in its little manual pump section. Your sketch should show the order in which they came out.

4. Are there wires to disconnect? Normally the small gas engine has only a few wire connections: from the coil to the spark plug, from the coil to the breaker points, and from the coil to the kill switch (engine stop). Your sketch should show where they go and how they are routed, and you should wrap a piece of masking tape around each and label the tape with a marking pen.

5. Is there tubing to disconnect? The typical small gas engine has many pieces of tubing; one chain saw has three pieces for the chain oiler alone. As with wiring, show how it's routed; then tape and label. Tubing is even more critical because the lengths are different and it's easier to confuse them during reassembly than you think.

Don't try to make do without the right tool. Yes, there are many small gas engine jobs in which you can do the work without a special tool, and this book carefully points them out. Don't take too many liberties, however, such as using pliers on a hex-head nut or bolt because you don't have the right size wrench, or trying to remove Phillips-head screws with a slot-type screwdriver, pliers or a Phillips-head driver that's too large or too small. These all are "no-noes," and the list is not exhaustive or it alone would fill a book. Sets of small socket wrenches, open-end and box wrenches, Allen wrenches and screwdrivers are inexpensive and have many uses in addition to work on small gas engines. If you don't have them, buy them and you'll find they pay for themselves in many ways.

If a part won't come off when all the retainers seem to be out, don't just start prying with a screwdriver. Take a second and third look, and you may find a screw hidden somewhere. On the other hand, just because all the retainers are out doesn't mean the part will fall out either. It may need some persuasion.

If a screw or bolt really is tight, as are flywheel nuts, starter drum screws, clutch retainers and a variety of other parts, no normal amount of force you can apply with an ordinary hand tool may break them loose. Professional shops use impact wrenches powered by compressed air or electricity. You can't justify an investment in these, but a household substitute is the manual impact tool, which comes with screwdriver bits and socket wrenches. You whack the top of this tool with a hammer and it works every bit as well as the professional equipment. If the screw or bolt is extremely recalcitrant, consider the possibility that it may have a reverse thread to the normal. This is occasionally seen on flywheel nuts, clutch retainers and starter drum screws. Just put the impact tool into the tighten position and try again.

Never disassemble any more than you have to. Most carburetors come out without removing any more than the nuts or bolts that hold them, and perhaps one cover plate. If you're replacing breaker points, just remove the cover plate that goes over the flywheel. The less you take apart, the less chance there is of making a mistake.

Don't panic. The worst that can happen is you'll have to shovel a driveway or let the lawn go for a week. Even professionals make mistakes, most commonly leaving out an important part or breaking a part. They simply accept the situation and start again. So should you.

Troubleshooting
Like a Pro

The worst way to waste time is to try to fix something that doesn't need fixing. Determining exactly what ails an engine is an exercise in logic called "troubleshooting." Within the equipment limits that even professional mechanics must work, there is always a certain amount of supposition and guesswork, but the good troubleshooter keeps his guesses to the minimum and tries to make the most effective check with the least expenditure of effort.

You must accept the fact that you will not be able to troubleshoot every conceivable problem. Every small gas appliance has some particular widget that can fail in some unusual way, and only the man who services these units on an everyday basis can hope to keep up with even a majority of the possibilities. What you can expect is to be able to find the routine causes of failure, the ones that account for 99 percent of the problems.

Although most of this chapter consists of troubleshooting charts, several of the popular items are discussed separately from the charts. You must understand that charts are merely memory jogs, to remind you to check something that you already know about. A chart can't find the trouble for you. If you see that "choke partly closed" is a possible cause for a problem, you must know how to check this possibility. To do this, you may have to do more than just pull the choke linkage to the open position. You should also know how to find the carburetor and know which plate in the air horn is the choke, in order to cover the possibility that something is wrong with the linkage or its adjustment.

You also must be able to tell when something's wrong by looking at it. A clogged air filter, for example, need not be black with dirt. If you've got sandy soil, that filter may look as if it just rolled off a production line, and yet be plugged. A fuel filter also can look clean—and actually be clean—but if it's water-logged because you didn't run the last bit of gasoline through when you packed away the machine for the season, it's just as plugged as if it were filled with dirt.

Can't tell? Not sure? You probably cannot take the time to master the trade of small gas engine service. All you want to be able to do is make most of your own repairs and keep the machine going as long as possible. There are times when you will have to take the appliance in for professional service, but if you keep cool and think logically, they will be few and far between.

When you're not sure, there are often simple little double checks, and this chapter explains them.

A sound piece of advice is to know your machine and not expect it to perform the way it did when it was new. This is not to say that hard starting and poor performance, stalling, vibration, or other problems are normal conditions that you must live with as a machine wears on. In many cases, careful service and maintenance will keep overall performance reasonably close to new-machine levels.

Note: Some of the items covered in the troubleshooting charts are not among the sort of things the average homeowner will want to tackle. They are included primarily to make you a somewhat more informed customer when you do have to bring the appliance to a service center. Here are some examples: You bring the appliance in for replacement of crankshaft bearings and soon after it's back in use, you notice oil leaks. The troubleshooting chart lists three things that could have been done wrong by the repair shop. Or let's say your engine has developed a performance problem and your checkout discloses that all the easy-to-service systems are in good condition. The extremely high probability that one of the major repairs on the list is required provides you with some of the information you need to make an "Is it worth fixing?" decision, covered in Chapter 9.

ENGINE WON'T START

The most common problem is that the engine fails to start, so let's begin with this one. There are literally a thousand possible causes of starting failure, but a few tests to check out the most frequently-occurring possibilities normally isolate the problem very quickly. The ingredients of a successful engine start are these:

1. The engine turns over (the job of the starter).
2. The piston develops adequate compression. (The engine must be in reasonably good mechanical condition.)
3. High-voltage current is delivered to the plug, and it jumps the air gap between the electrodes.

4. The spark arrives at the right time.
5. A mixture of air and fuel in reasonably proper proportions is delivered to the cylinder.

QUICK CHECKS

A series of quick checks will isolate the problem to one of the five items. It may not give you the solution, but it's the only way to begin.

When you pull the starter cord you can feel and hear if the engine is turning. If it isn't, the problem is the starter. (Refer to Chapter 6.)

Disconnect the spark plug wire and hold it ⅛ to ¼-inch from the spark plug terminal. (If the plug wire has a rubber boot, wedge a paper clip or coil spring into the metal connector it covers; make sure you use a clip or spring that is long enough to project from the boot.) See 4–1. Crank the engine and you should see a spark jump to the plug at least a couple of times during each complete pull of the starter cord. If you have an electric starter, the wire should discharge a spark at regular intervals for as long as the starter operates. If you get the sparks, the ignition system up to the plug is in good condition. If you don't, check the troubleshooting chart; then review ignition system service as explained in Chapter 6.

Caution: If a spark plug wire's insulation is defective, holding the wire with your bare hands could lead to an electrical shock. To avoid this danger, hold the wire with a pair of small sticks, chopstick fashion.

With a spark to the plug, the next check begins with leaving the plug wire off and cranking the engine several times. Now remove the plug

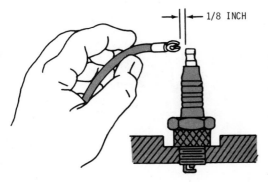

→ ‖ ← 1/8 INCH

4–1. You can perform a quick check of the ignition system by holding the plug wire about ⅛ inch from the plug's center electrode while the engine is being cranked. The spark should jump from the wire terminal to the plug electrode if the ignition system is okay.

quickly and inspect it. If the fuel system is delivering, the plug will be wet with gasoline. If it isn't, you have a fuel system problem. You can double-check a no-fuel problem by removing the spark plug, pouring a couple of teaspoons of gas through the hole into the cylinder, reinstalling the plug and trying to start the engine. If the engine now starts and runs for an instant, lack of fuel is the problem, caused by anything from an empty tank to a plugged fuel line. Note: On chain saws, use a half teaspoon of fuel-oil mix. Also inspect the spark plug's general condition. If you have any doubt about it, reconnect the plug wire and using rubber or an insulated pliers as a holder, rest the plug against the cylinder head while a helper cranks the engine. You should see the spark jump the gap across the electrodes. See 4–2. If you had a satisfactory spark up to the plug and none jumping the gap, the plug is fouled and should be replaced.

If fuel is wetting the plug, you know that some amount is being delivered, but you don't know if it's too much. To check out this possibility (flooding), open the choke and the throttle all the way; then try to start the engine. If the engine now gives at least some indication of firing (or if it starts and runs), flooding is indicated. A strong odor of gasoline at the carburetor is a confirming clue. These problems are covered in Chapter 7.

The next check is for adequate compression. This can be done with an automotive compression gauge on those engines for which compression specifications are published. Unlike an automobile engine, however, the

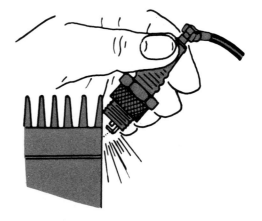

4–2. To check the spark plug itself, remove it and hold it against the cylinder head as shown, with the L-shaped side electrode in the base touching the head to complete a circuit to ground. Crank the engine and, if the plug is in good condition, the spark should jump from the center electrode to the L-shaped side electrode. If the ignition system checked out as shown in 4–1, but the spark does not appear, you should replace the plug.

small gas engine compression reading is a very rough indication. In fact, Briggs and Stratton refuses to publish compression specifications on the theory that too many people regard specifications as gospel, that mistakes are easily made taking the readings, and that unnecessary repairs result.

Inasmuch as Briggs is Number 1, this philosophy cannot be ignored. It suggests that you pull the flywheel cover and spin the flywheel counterclockwise against the compression stroke (spark plug reinserted and tight). You can identify this stroke by turning the flywheel until you encounter resistance, which will be the piston pressing against the air-fuel mixture. Give the flywheel a flip motion counterclockwise, and it should rebound very sharply. A slight rebound or none at all indicates poor compression.

If the manufacturer specifies a compression reading, have a helper hold the compression gauge firmly in the plug hole while you crank the engine with the rope starter half a dozen times. See 4–3. Most small gas engines will read at least 50 to 60 pounds, although those with a compression release feature (for easier starting) might read 40 to 45 pounds. These are not

See specifications
for correct
compression

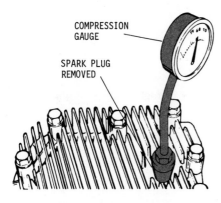

COMPRESSION
GAUGE

SPARK PLUG
REMOVED

4–3. If you have compression specifications, you can perform a compression test as shown in this illustration. If the compression reading (in psi) meets specifications, the engine is in mechanically acceptable condition. The purpose of repeating the compression test with oil in the cylinder is to test the piston rings. If the compression reading rises substantially (by one-third or more) with oil, the rings are worn. The oil provides a temporary seal for the piston. hence the increase in the reading. Although not likely, it is possible for an engine to meet minimum compression specifications with badly worn rings and fail to pass the second part of the test. Such an engine still is in need of overhaul.

FIRST TEST: Without oil in cylinder
SECOND TEST: Squirt a few drops of oil onto piston
              through spark plug hole
EACH TEST: Turn engine over 6 to 8 revolutions

(Note: For compression-release models,
 reading will be lower.)

universal readings. For the McCulloch MAC-10 series, a minimum reading would be 100 pounds. It's a good idea to make this test on your engine when it's new, so that you know what it should be, then allow a 25 percent drop from normal engine wear.

Low compression normally is caused by 1) a loose spark plug, 2) a defective cylinder head gasket, 3) valves that are burned or failing to close, or 4) worn or stuck piston rings. Unless the starting problem occurs at the start of a season, when aging and poor storage could have caused the valves or piston rings to stick, the loose plug or cylinder head gasket is the most likely possibility if the engine ran normally the last time it was used. A head gasket failure can occur suddenly, and when it does, the loss of power is instantaneous. If the engine very suddenly ran poorly during the last use, bet on the head gasket or the spark plug. Poor performance is not a certain indication of a head gasket failure though, for the gasket could rupture as the engine is being shut down.

If the fuel, ignition, compression and starter all check out, ignition timing is suspect by process of elimination. On an engine where coil position controls timing, take a look at the magneto coil to see if the screws are holding it tight. On an engine whose plate position adjusts timing, remove the flywheel and check the breaker plate bolts with a wrench, even slightly loose bolts can allow an unwanted timing change.

ENGINE IS HARD TO START

When the engine is difficult to start, but finally does get going, the most common causes are usually similar to those that cause complete starting failures, namely:

1. Partly clogged fuel line, fuel filter or fuel system vent hole (usually in gas tank cap)
2. Low compression
3. Fouled spark plug
4. Worn breaker points
5. Loose connection at magneto or points
6. Deteriorated spark plug wire
7. Choke's failure to close completely
8. Defective but not completely failed magneto coil
9. Water in gasoline.

Conditions 4, 5, 6, and 8 will result in a weak spark, one that will jump across the electrodes in the plug or to an electrical ground intermittently. Number 1 will show a virtually dry spark plug. Number 9 is a likely possibility if you didn't drain the tank and refill with fresh gasoline at the start of the season. As an extreme possibility, water in the gas tank might result in what seems to be a complete starting failure if you happen to give up a bit early.

The possibility of a plugged fuel system vent deserves some explanation. When the fuel pump or engine vacuum draws gasoline out of the tank, air must fill the gap. If the vent (usually a pinhole in the gas tank cap) is plugged, then air cannot enter. Result: There is a vacuum in the gas tank that keeps the fuel there. If the vent is partly clogged, only a limited amount of air can replace the fuel, and therefore only a limited amount of fuel can be drawn into the carburetor, causing the engine to suffer from fuel starvation. To check for this possibility, try to start the engine with gas cap removed.

ENGINE STALLS

The most common cause of engine stalling is a clogged air filter (something you can test by simply trying to start the engine with the filter off). Another is failure to open the choke once the engine has warmed up. There are other causes, however, such as the plugged fuel system vent and water in the gasoline, and many others.

ENGINE VIBRATES

Many people panic when the engine vibrates, assuming the worst. In most lawn mower cases the problem isn't in the engine at all; here the blade is probably bent or damaged. Another common cause for vibration is that the engine mounting bolts are loose. Still another is a worn flywheel retaining key. See the appropriate troubleshooting chart at the end of this chapter for all the common possibilities.

POOR PERFORMANCE

The chart on page 62 shows many common internal engine causes of poor performance, but here are others you can check for:

1. Choke partly closed, or clogged air filter causing an overrich mixture
2. Poor compression
3. Inadequate lubrication
4. Dirty carburetor
5. Retarded ignition timing: The combustion pattern is altered, reducing the completeness of the combustion process.
6. Improperly adjusted carburetor: This is most likely to occur after the carburetor has been rebuilt.
7. Engine running too hot: The most common reason for a small-gas-engine's overheating is that the air flow is restricted by leaves, wood shavings or ice; the cause and equipment relationship is obvious.
8. Mower or blower air vane governor packed with leaves, preventing the throttle from opening

ENGINE OVERHEATS

After a restricted air flow, the second-most common cause of engine over-heating (in my neighborhood anyway) is a 250-pound homeowner, who thought he got a bargain, trying to get his 5-horsepower riding mower to cut through thick grass on an upgrade. What this comes down to is: Don't overload the equipment. A little snow blower, or a mower with a snow-throwing adapter can't handle a three-foot snow, and a tiny saw with a dull chain is not going to cut down a big tree, no matter how much push you give.

Overadvanced ignition timing is another cause of overheating. Advancing the ignition timing beyond manufacturer's specifications (30 degrees before top-dead-center instead of 26, as an example) may give you a tad more power, but the combustion chamber temperatures will rise so much that the engine will overheat. Each engine is an individual in this respect, so the best policy is to stick with manufacturer's specifications.

2-CYCLE CHAIN SAW TROUBLESHOOTING CHART

Engine will not start, or is hard to start, or stalls or misfires. (Numbers 1 through 14 are fuel system problems; 15 through 23 ignition system problems; 24 through 31 mechanical problems)

Possible Cause	Remedy
1. Out of fuel or water in fuel	Drain tank and blow out fuel lines to remove water; refuel tank.
2. Wrong mixture	Adjust mixture screws.
3. Clogged or waterlogged fuel filter	Replace fuel filter.
4. Plugged air filter	Clean or replace air filter.
5. Clogged fuel tank vent	Clean tank vent.
6. Plugged fuel line	Blow out fuel line.
7. Cut or leaking fuel-filter fitting in tank (on saws so equipped)	Replace filter fitting.
8. Fuel-pump diaphragm defective	Replace pump diaphragm.
9. Fuel-pump filter screen plugged	Clean filter screen or replace.
10. Fuel pump passage clogged	Clean out carburetor.
11. Carburetor inlet flow-control needle valve worn	Replace needle valve assembly.
12. Carburetor jet needles worn or improperly adjusted	Adjust or replace needles.

Possible Cause	Remedy
13. Carburetor gaskets leaking	Replace carburetor gaskets.
14. Carburetor diaphragm defective	Replace diaphragm.
15. Spark plug fouled, improperly gapped or broken	Replace or regap spark plug.
16. Magneto kill switch grounding out system even when in "on" position	Adjust switch tab if possible; otherwise replace switch.
17. Kill switch in "off" position	Reposition switch knob.
18. Condenser defective or ground wire connection loose	Replace condenser or tighten connection.
19. Breaker points burned, dirty, improperly gapped, or fixed point not grounded.	Replace or regap breaker points; or check for poor ground connections and tighten or repair as necessary. (If a separate grounding wire is used on breaker plate, check its connections.)
20. Ignition timing incorrectly set	Reset ignition timing.
21. Air gap between flywheel and magneto coil improperly set	Readjust magneto coil position with feeler gauge.
22. Magneto wiring to breaker points or spark plug defective or broken	Repair wiring to breaker points; or replace spark plug wire.
23. Magneto coil defective	Replace magneto coil.
24. Crankcase bolts or nuts loose	Tighten crankcase bolts or nuts.
25. Piston rings worn	Overhaul engine.
26. Flywheel half-moon key sheared	Replace half-moon key.
27. Reed valve defective	Replace reed valve.
28. Cylinder cracked	Replace top half of engine.
29. Hole in piston	Replace piston.
30. Crankcase seals leaking	Replace crankcase seals.
31. Leaking reed valves	Replace reed valves.

Engine Performance Is Poor, or Engine Runs Only When Choked (Numbers 1 through 3 are fuel system problems; 4 an ignition system problem; 5 through 11 mechanical problems.)

Possible Cause	Remedy
1. Air filter clogged	Clean or replace air filter.
2. Fuel filter clogged or water-logged	Replace fuel filter.
3. Carburetor needles out of adjustment	Adjust mixture screws.

2-CYCLE CHAIN SAW: PERFORMANCE (Continued)

Possible Cause	Remedy
4. Ignition timing incorrectly set	Reset ignition timing.
5. Piston rings worn	Overhaul engine.
6. Muffler plugged	Replace muffler.
7. Exhaust ports plugged	Clean carbon from ports.
8. Crankcase seals leaking	Replace crankcase seals.
9. Piston and/or cylinder scored	Replace piston and top half of engine if necessary.
10. Air leak at carburetor base gasket	Tighten carburetor mounting nuts or bolts or replace carburetor base gasket.
11. Leaking reed valves	Replace reed valves.

Engine won't run at full speed. (Numbers 1 through 7 are fuel system problems; 8 through 9 ignition system problems; 10 through 13 mechanical problems.)

Possible Cause	Remedy
1. Air filter clogged	Clean or replace air filter.
2. Carburetor diaphragm defective	Replace diaphragm.
3. Carburetor inlet needle valve dirty or defective	Replace needle valve assembly.
4. Gas tank vent partly plugged	Clean tank vent.
5. Throttle plate not opening fully	Loosen throttle plate screws and reposition throttle.
6. Carburetor needles out of adjustment	Adjust mixture screws.
7. Fuel filter partly plugged or waterlogged	Replace fuel filter.
8. Ignition timing incorrectly set	Reset ignition timing.
9. Spark plug defective (breaking down at high speed)	Replace spark plug.
10. Muffler partly plugged	Replace muffler.
11. Exhaust ports partly plugged	Clean carbon from ports.
12. Rings worn	Overhaul engine.
13. Leaking reed valves	Replace reed valves.

Engine overheats. (Numbers 1 through 4 are fuel system problems; 5 and 6 ignition system problems; 7 through 11 mechanical problems.)

Possible Cause	Remedy
1. Improperly set carburetor needles	Adjust mixture screws.
2. Air leak at carburetor base gasket	Tighten carburetor mounting nuts or bolts or replace carburetor base gasket.
3. Carburetor internal defect	Overhaul carburetor.
4. Dirty air filter	Clean or replace air filter.
5. Incorrect spark plug	Replace spark plug with correct part.
6. Ignition timing improperly set	Reset ignition timing.
7. Sawdust guard dirty or plugged	Clean sawdust guard.
8. Cylinder fins dirty	Clean cylinder fins.
9. Flywheel vanes broken	Replace flywheel.
10. Flywheel cover housing loose	Tighten flywheel cover housing.
11. Muffler plugged	Replace muffler.

Engine starts but stops after running briefly.

Possible Cause	Remedy
Fuel tank vent partly plugged	Clean tank vent.
Water in fuel mixture	Drain tank and blow out fuel lines.
Air filter clogged	Clean or replace air filter.
Carburetor inlet needle valve or passages dirty	Replace needle valve assembly or clean out carburetor.
Carburetor diaphragm defective	Replace diaphragm.
Air leak at carburetor base gasket	Tighten carburetor mounting nuts or bolts, or replace carburetor base gasket.

Engine starves on acceleration.

Possible Cause	Remedy
Air filter clogged	Clean or replace air filter.
Fuel filter waterlogged or clogged	Replace fuel filter.

2-CYCLE CHAIN SAW: ACCELERATION (Continued)

Possible Cause	*Remedy*
Carburetor needles improperly adjusted	Adjust mixture screws.
Carburetor inlet valve defective	Replace inlet needle valve assembly.
Carburetor inlet control lever (if used) bent	Bend lever back if possible, otherwise replace lever.

Clutch slips (under load).

Possible Cause	*Remedy*
Shoes worn or stuck	Replace shoes, and drum if necessary.
Chain too tight on guide bar	Adjust chain.
Chain improperly filed, causing drag	File chain correctly or replace.

Clutch drags or rattles.

Possible Cause	*Remedy*
Shoes stuck or worn	Replace shoes, and drum if necessary.
Spring weak or broken	Replace spring if available as separate part; otherwise replace shoes and spring assembly.
Drum out of round	Replace drum.
Sprocket bearing defective	Replace sprocket bearing.

Chain oiler doesn't work.

Possible Cause	*Remedy*
Reservoir empty	Refill reservoir.
Oil too heavy	Mix in kerosene to thin out oil sufficiently to drain properly; then refill with correct thickness oil.
Internal pump defect (dirt, worn seals, worn plunger, defective diaphragm if used, defective check valve, worn plunger bore)	Replace or rebuild pump assembly.

Possible Cause	Remedy
Pump outlet line or neck plugged	Clean or replace pump outlet line
Reservoir cap vent plugged	Clean reservoir cap vent.
Reservoir pickup tube or screen clogged	Clean or replace reservoir pickup tube or screen.

4-CYCLE ENGINE TROUBLESHOOTING CHART

Engine Fails to Start, Starts with Difficulty, or Runs Only When Choked

Possible Cause	Possible Remedy
No fuel in tank	Fill tank with clean, fresh fuel.
Shut-off valve closed	Open valve.
Obstructed fuel line	Clean fuel screen and line. If necessary, remove and clean carburetor.
Tank cap vent obstructed	Open vent in fuel tank cap.
Water in fuel	Drain tank. Clean carburetor and fuel lines. Dry spark plug electrodes. Fill tank with clean, fresh fuel.
Engine over-choked	Close fuel shut-off and pull starter until engine starts. Reopen fuel shut-off for normal fuel flow.
Improper carburetor adjustment	Adjust carburetor.
Loose or defective magneto wiring	Check magneto wiring for shorts or grounds; repair if necessary.
Faulty magneto	Check timing, point gap, and, if necessary, replace magneto.
Spark plug fouled	Replace spark plug.
Spark plug porcelain cracked	Replace spark plug.
Poor compression—leaking valves	Perform valve job:
Poor compression because of worn piston rings	Overhaul engine.
Defective magneto kill switch	Replace kill switch.

4-CYCLE ENGINE (Continued)

Engine Knocks

Possible Cause	Possible Remedy
Carbon in combustion chamber	Remove cylinder head or cylinder and clean carbon from head and piston.
Loose or worn connecting rod	Replace connecting rod and bearings.
Loose flywheel	
	Check flywheel key and keyway; replace parts if necessary. Tighten flywheel nut to proper torque.
Worn cylinder	Replace cylinder.
Improper magneto timing	Time magneto.
Excessive crankshaft end play	Replace main bearings.

Engine Misses Under Load

Possible Cause	Remedy
Spark plug fouled	Replace spark plug.
Spark plug porcelain cracked	Replace spark plug.
Improper spark plug gap	Regap spark plug
Pitted magneto breaker points	Replace pitted breaker points.
Points' breaker arm sluggish	Clean and lubricate breaker point arm pivot.
Faulty condenser	Replace condenser.
Improper carburetor adjustment	Adjust carburetor.
Improper valve clearance	Adjust valve clearance to specifications.
Weak valve spring	Replace valve spring.

Engine Vibrates Excessively

Possible Cause	Remedy
Engine not securely mounted	Tighten loose mounting bolts.
Bent crankshaft	Replace crankshaft.
Associated equipment out of balance	Check associated equipment.

Breather Passing Oil

Possible Cause	Remedy
Engine speed too fast	Use tachometer to adjust correct rpm. Check governor and spring.
Loose oil fill cap or gasket damaged or missing	Install new ring gasket under cap and tighten securely.
Oil level too high	Check oil level: Turn dipstick cap tightly into receptacle for accurate level reading. DO NOT fill above full mark.
Breather mechanism damaged	Check valve assembly. Replace complete unit if necessary.
Breather mechanism dirty	Clean thoroughly in solvent. Use new gaskets when reinstalling unit.
Drain hole in breather box clogged	Clean hole with wire to allow oil to return to crankcase.
Piston ring end gaps aligned	Rotate end gaps so as to be staggered 90° apart.
Breather mechanism installed upside down	Small oil drain holes must be down to drain oil from mechanism.
Breather mechanism loose or gaskets leaking	Install new gaskets and tighten securely.
Damaged or worn oil seals on end of crankshaft	Replace seals.
Rings not seated properly	Check for worn or out of round cylinder. Replace rings. Break in new rings with engine working under a varying load. Rings must be seated under high compression or in other words under varied load conditions.
Breather assembly not assembled correctly	Check assembly; reconnect as necessary.
Cylinder cover gasket leaking	Replace cover gasket.

Engine Lacks Power

Possible Cause	Remedy
Ignition system malfunction	Check plug, points, condenser, wiring, magneto.
Magneto improperly timed	Time magneto.
Choke partially closed	Open choke.

4-CYCLE ENGINE: LACKS POWER (Continued)

Possible Cause	Remedy
Improper carburetor adjustment	Adjust carburetor.
Worn rings (low compression)	Replace rings.
Lack of lubrication	Fill crankcase to the proper level.
Air cleaner fouled	Clean air cleaner.
Valves leaking (low compression)	Perform valve job.
Internal carburetor defect	Clean and rebuild carburetor.

Engine Overheats

Possible Cause	Remedy
Engine improperly timed	Time engine.
Carburetor improperly adjusted	Adjust carburetor.
Air flow obstructed	Remove any obstructions from air passages in shrouds.
Cooling fins clogged	Clean cooling fins.
Excessive load on engine	Check operation of associated equipment. Reduce excessive load.
Carbon in combustion chamber	Remove cylinder head or cylinder and clean carbon from head and piston.
Lack of lubrication	Fill crankcase to proper level.

Engine Surges or Runs Unevenly

Possible Cause	Remedy
Fuel tank cap vent hole clogged	Open vent hole.
Governor parts sticking or binding	Clean, and if necessary repair governor parts.
Carburetor throttle linkage or throttle shaft and/or butterfly binding or sticking	Clean, lubricate, or adjust linkage and deburr throttle shaft or butterfly.
Governor malfunctioning	Adjust or repair as necessary.
Carburetor adjusted improperly	Adjust to specifications.

Engine Uses Excessive Amount of Oil

Possible Cause	*Remedy*
Engine speed is too fast.	Using tachometer adjust engine rpm to specifications.
Oil level is too high.	Check level: Turn dipstick cap tightly into receptacle for accurate level reading.
Oil filler cap is loose or gasket is damaged causing spillage out of breather.	Replace ring gasket under cap and tighten cap securely.
Breather mechanism is damaged or dirty causing leakage.	Replace breather assembly.
Drain hole in breather box is clogged causing oil to spill out of breather.	Clean hole with wire to allow oil to return to crankcase.
Gaskets are damaged or gasket surfaces are nicked, causing oil to leak out.	Clean and smooth gasket surfaces. Always use new gaskets.
Valve guides worn excessively thus passing oil into combustion chamber.	Ream valve guide oversize and install oversize valve.
Cylinder wall worn or glazed allowing oil to pass rings into combustion chamber.	Bore or deglaze cylinder as necessary.
Piston rings and grooves worn excessively.	Install new rings and check land clearance and correct as necessary.
Piston fit undersized.	Measure and replace as necessary.
Piston oil control ring return holes clogged.	Remove oil control ring and clean return holes.
Oil passages obstructed.	Clean out all oil passages.

Oil Seal Leaks

Possible Cause	*Remedy*
Old seal hardens and is worn.	Replace old, hardened seal.
Crankshaft seal contact surface is slightly scratched causing seal to wear excessively.	Crankshaft seal rubbing surface must be smoothed before installing new seal. Use a fine crocus cloth. Care must be taken when removing seals.
Crankshaft seal contact surface is worn undersize causing seal to leak.	Check crankshaft size and replace if worn excessively.
Crankshaft bearing under seal is worn excessively causing crankshaft to wobble in oil seal.	Check crankshaft bearings for wear and replace if necessary.

Possible Cause	Remedy
Seal outside seat in cylinder or side cover is damaged allowing oil to seep around outer edge of seal.	Visually check seal receptacle for nicks and damage. Replace power takeoff cylinder cover or small cylinder cover on the magneto end, if necessary.
New seal installed without correct seal driver and not seating squarely in cavity.	Replace with new seal using proper tools and methods.
New seal damaged upon installation.	Use proper seal protector tools and methods for installing another new seal.
Bent crankshaft causing seal to leak.	Check crankshaft for straightness and replace if necessary.
Oil seal driven too far into cavity.	Remove seal and replace with new seal using the correct driver tool and procedures.

TROUBLESHOOTING CARBURETION

Trouble	*Corrections (Keyed to Next Page)*
Carburetor out of adjustment	**3−11−12−13−15−20**
Engine will not start	**1−2−3−4−5−6−8−11−1 2−14−15−24−25**
Engine will not accelerate	**2−3−11−12−24**
Engine hunts (at idle or high speed)	**3−4−8−9−10−11−12−1 4−20−21−24−26**
Engine will not idle	**4−8−9−11−12−13−14− 20−21−22−24−25−26**
Engine lacks power at high speed	**2−3−6−8−11−12−20−2 1−24−25−26**
Carburetor floods	**4−7−17−21−22−25−26**
Carburetor leaks	**6−7−10−18−23−24**
Engine overspeeds	**8−9−11−14−15−18−20**
Idle speed is excessive	**8−9−13−14−15−18−20−25−26**
Choke does not open fully	**8−9−15**
Engine starves for fuel at high speed (leans out)	**1−3−4−6−11−15−17−21−26**
Carburetor runs rich with main adjustment needle shut off	**7−11−17−18−19−21−25−26**
Performance unsatisfactory after being serviced.	**1−2−3−4−5−6−7−8−9−10−11−15−16−17−18−20−21−25−26**

Corrections (Keyed to Previous Page)

1. Open fuel shut-off valve at fuel tank; fill tank with fuel.

2. Check ignition, spark plug and compression.

3. Clean air cleaner; service as required.

4. Dirt or restriction in fuel system: Clean tank and fuel strainers, check for kinks or sharp bends.

5. Check for stale fuel or water in fuel; fill with fresh fuel.

6. Examine fuel line and pickup for sealing at fittings.

7. Check and clean atmospheric vent holes.

8. Examine throttle and choke shafts for binding or excessive play; remove all dirt or paint, replace shaft.

9. Examine throttle and choke return springs for operation.

10. Examine idle and main mixture adjustment screws and "O" rings for cracks or damage.

11. Adjust main mixture adjustment screw; some models require finger tight adjustment. Check to see that it is the correct screw.

12. Adjust idle mixture adjustment screw. Check to see that it is the correct screw.

13. Adjust idle speed screw.

14. Check for bending choke and throttle plates.

15. Adjust control cable or linkage to assure full choke and carburetor control.

16. Clean carburetor after removing all nonmetallic parts that are serviceable. Trace all passages.

17. Check inlet needle and seat for condition and proper installation.

18. Check sealing of welch plugs, cups, plugs and gasketes.

20. Check governor linkage; clean if necessary.

Specific Carburetor Checks for Float

21. Adjust float setting.

22. Check float shaft for wear and float for leaks or dents.

23. Check seal for fuel drain or bowl gasket.

24. Is carburetor operating at excessive angle—31° or more?

Specific Carburetor Checks for Diaphragm

25. Check diaphragm for cracks or distortion. If nylon check ball is present, check for function.

26. Check sequence of gasket and diaphragm for the particular carburetor being repaired.

POINTS TO CHECK FOR ENGINE POWER

Ignition: Must be properly timed so that spark plug fires at precise moment for full power.

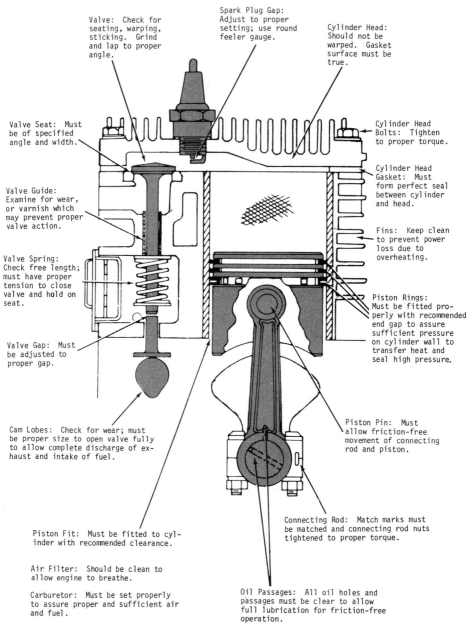

Valve: Check for seating, warping, sticking. Grind and lap to proper angle.

Spark Plug Gap: Adjust to proper setting; use round feeler gauge.

Cylinder Head: Should not be warped. Gasket surface must be true.

Valve Seat: Must be of specified angle and width.

Cylinder Head Bolts: Tighten to proper torque.

Cylinder Head Gasket: Must form perfect seal between cylinder and head.

Valve Guide: Examine for wear, or varnish which may prevent proper valve action.

Fins: Keep clean to prevent power loss due to overheating.

Valve Spring: Check free length; must have proper tension to close valve and hold on seat.

Piston Rings: Must be fitted properly with recommended end gap to assure sufficient pressure on cylinder wall to transfer heat and seal high pressure.

Valve Gap: Must be adjusted to proper gap.

Cam Lobes: Check for wear; must be proper size to open valve fully to allow complete discharge of exhaust and intake of fuel.

Piston Pin: Must allow friction-free movement of connecting rod and piston.

Connecting Rod: Match marks must be matched and connecting rod nuts tightened to proper torque.

Piston Fit: Must be fitted to cylinder with recommended clearance.

Air Filter: Should be clean to allow engine to breathe.

Carburetor: Must be set properly to assure proper and sufficient air and fuel.

Oil Passages: All oil holes and passages must be clear to allow full lubrication for friction-free operation.

62

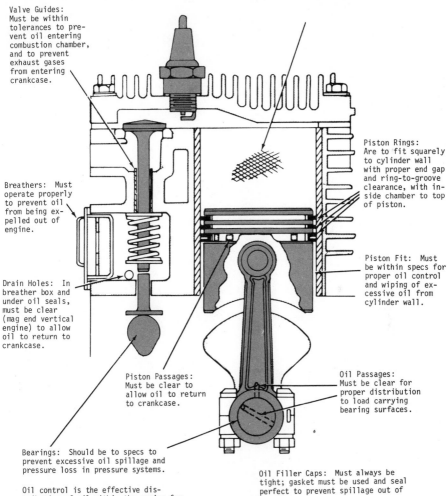

Cylinder Wall Finish: Cylinder wall glaze must be broken prior to installing new piston rings to allow rings to seat and control oil.

Valve Guides: Must be within tolerances to prevent oil entering combustion chamber, and to prevent exhaust gases from entering crankcase.

Piston Rings: Are to fit squarely to cylinder wall with proper end gap and ring-to-groove clearance, with inside chamber to top of piston.

Breathers: Must operate properly to prevent oil from being expelled out of engine.

Piston Fit: Must be within specs for proper oil control and wiping of excessive oil from cylinder wall.

Drain Holes: In breather box and under oil seals, must be clear (mag end vertical engine) to allow oil to return to crankcase.

Piston Passages: Must be clear to allow oil to return to crankcase.

Oil Passages: Must be clear for proper distribution to load carrying bearing surfaces.

Bearings: Should be to specs to prevent excessive oil spillage and pressure loss in pressure systems.

Oil control is the effective distribution of oil within the engine for friction-free operation and the prevention of oil being burned or leaked at gasket surfaces.

Gasket Surfaces: Must not be nicked; when old gaskets are removed, always install new gaskets.

Oil Filler Caps: Must always be tight; gasket must be used and seal perfect to prevent spillage out of breather.

Oil Level: Overfilling will cause leaking, burning and oil-fouled spark plugs.

Engine Speed: Excessive speeds will cause excessive oil comsumption by burning and leaking.

Servicing Engines

Removing a small gas engine from its chassis and taking it apart sufficiently for a majority of repairs is well within the ability of the average homeowner. This chapter stops short of engine overhaul, but includes illustrations to explain some of the complications. As an experiment, some very experienced small-gas-engine repairmen tried to perform engine overhaul without special shop equipment, and even they (knowing the difficulties) ran into problems, including broken piston rings.

If you limit yourself to such jobs as ignition service, fuel system work, clutch replacement, chain saw pump oiler rebuilding and engine jobs as covered in this chapter, and make good maintenance a part of your routine, the engine overhaul will not come up for years.

ENGINE REMOVAL

Removing the engine is a job you wouldn't attempt on a car unless absolutely necessary, but on most mowers and blowers the job is quick and easy, and an engine alone on a work bench provides an optimum working situation. See 5–1 through 21.

The only small gas engine exception to the take-it-out approach is the chain saw. On some chain saws the engine is part of the entire housing, not surprising in homeowner models where the manufacturer is putting the biggest power package possible in a compact container. Even if you can remove the chain saw engine, the job generally requires a fair bit of cover and component removal. So when you're working on this type of small gas appliance, try to do as many repairs as possible without a complete teardown. As you'll see in the illustrated disassembly sequences later in this chapter, you can get to most of the parts on a chain saw without taking everything apart.

TIGHTENING SPECIFICATIONS

Before you disassemble major components of a small gas engine, you should know the retightening specifications, which are given in pounds feet or pounds inches. (Divide by 12 to convert pounds inches to pounds feet and round off to the nearest whole number.) Tightening of certain nuts and bolts must be reasonably precise, and the only way to do this is with a regular torque wrench. The wrench has a dial and a needle that indicate pressure in foot-pounds on a scale of from 0 to 100. This is more than enough for a small gas engine, but if you plan to do any work on a car, you need a 0–150 scale. The 0–100 wrench costs about $10. The 0–150 wrench runs a few dollars more.

Before removing a bolt or nut you will later have to tighten, check the existing tightness. So if you don't have specifications, at least you can get an idea. To measure tightness, apply pressure with the torque wrench until the nut or bolt is just ready to move. Read the scale and you've got the original setting anyway. Because a nut or bolt may loosen (or rust and become tighter), the best time to check tightness is when the machine is new. You might even discover a factory goof—a nut or bolt that's ready to come off. You might also discover unevenness in tightening. Example: All head bolts are 17 pounds feet except one, which is as low as 8 or perhaps as high as 25. The nuts or bolts that should be torqued to specifications (as opposed to being just tight) are the cylinder head bolts, the flywheel nut, and the nuts or bolts that hold the crankcase and cylinder together.

REMOVING A VERTICAL CRANKSHAFT ENGINE

The vertical crankshaft engine on the typical rotary mower lifts right out after removal of the blade and the four nuts and bolts that hold it to the chassis. See 5–1 through 5–3.

If the mower is self-propelled, you must also disconnect the chain or belt that drives the wheels. In most cases all you do is push the engine toward the driven wheels (to the front if the front wheels are propelled, to the rear if the rear wheels are driven). This will provide enough slack in the belt or chain so you can lift it off. See 5–4 and 5.

A slightly different wrinkle is used on the Jacobsen self-propelled mower, as shown in 5–6 through 14. A pulley adapter is attached to the crankshaft between the centrifugal clutch and the blade, and a belt runs from this pulley to a pulley on a small gearbox at the front of the mower. A third pulley, the idler, is mounted under the engine and is spring-loaded to keep the belt tight. When you remove the engine mounting bolts and blade you can push the idler pulley forward to relieve spring tension, then slip the belt off. Remove an engine mounting bolt spacer (5–13) and the belt can be pushed through the opening and removed.

5–1 and 5–2. Engine removal on the simple rotary mower begins with removal of the blade. The exact procedure varies according to mower, but it is something your owner's manual should cover if it is not completely obvious. You may have to hold the blade with one hand and whack the end of the wrench with a hammer to break the nut loose.

5–3. Once the blade is off, remove all the bolts, usually four, that hold the engine to the chassis.

5—4. If the mower is self-propelled, there may be a drive chain to disconnect. In this case, the engine bolts will pass through elongated holes in the chassis, permitting you to slide the engine forward far enough to lift the chain off its sprocket.

5—5. Then just lift the engine up and out, and take it to the workbench.

5—6. On this Jacobsen mower, engine removal begins with removal of the side cover, held by Phillips-head screws.

5—7. Blade removal on the Jacobsen is straightforward. Just loosen the bolt with a wrench and then remove it.

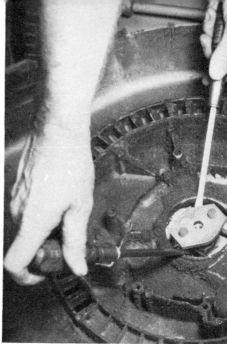

5-8 and 5-9. Once the blade is off, left, you can see the adapter hub that holds it, and part of the rubber belt that drives it. Then pry off the adapter hub with two screwdrivers as shown.

5-10. Here are the hub and the half-moon key that holds the hub. Don't lose or forget the key.

5-11 and 5-12. If it's necessary to change the Jacobsen belt, pull the pulley under the engine forward, against the spring, to release tension on the belt, as shown left. Then you can slip the belt off the pulley, which is then pulled back by a spring.

5-13. The belt still can't be removed because the engine mounting bolt spacer (to which pen points) blocks the way.

5-14. Remove the engine mounting bolt from underneath, here noted also by the pen point. This will permit you to push the spacer aside and change the belt. If you are taking out the engine, you can leave the belt in place and simply remove all engine mounting bolts and pull the engine, leaving the belt on the chassis. Just disengage the belt from the crankshaft under the blade, remove the engine mounting bolts and lift the engine out.

REMOVING A HORIZONTAL CRANKSHAFT ENGINE

The horizontal crankshaft engine, normally used on blowers and reel mowers, also is held by four bolts. A chain or belt drives the reels or auger in front. Moving the engine forward provides sufficient slack so that you can disconnect the chain or belt. See 5–15 through 21.

VALVE JOB (4-STROKE CYCLE ENGINE)

The valve job (5–22 through 46) is one major repair that a homeowner can tackle with some confidence after he's gotten some experience with other small gas engine repairs. It sounds much more difficult than it is, so if you proceed slowly and carefully, you're sure to have the job turn out successfully. Following is a complete beginner's guide to the valve job.

First, let's talk about what the job entails and why it's done.

A valve job, commonly called a carbon and valve job, begins with removing the cylinder head and the valve cover, often called the tappet cover, named after a little cylindrical part that sits on top of the cam lobe

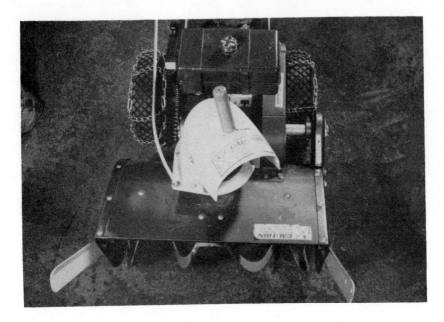

5-15. This self-propelled snow blower's engine comes out a bit differently from that of a mower, but it's still an easy job.

5-16 and 17. Begin removal of the engine by taking out the bolt and nuts that hold the chain cover, as shown left. Then the chain cover lifts away, exposing the chain.

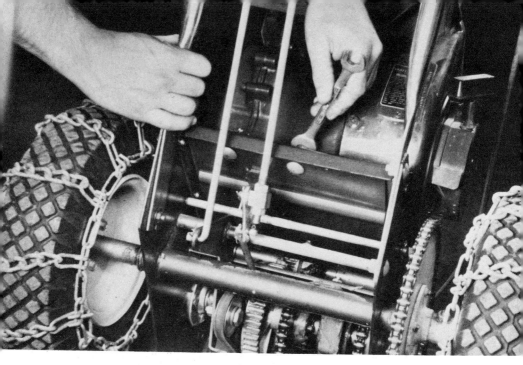

5–18. Next remove the bolts that hold the snow blower engine to the chassis. There usually are four.

5–19. After you slide the engine forward, you'll have enough slack to disengage the chain from the engine sprocket.

5–20. Lift the engine up and out.

5–21. With the engine out you can see the elongated holes in the chassis very clearly. When the engine is reinstalled, it is pushed rearward (with the chain refitted) to properly tension the chain.

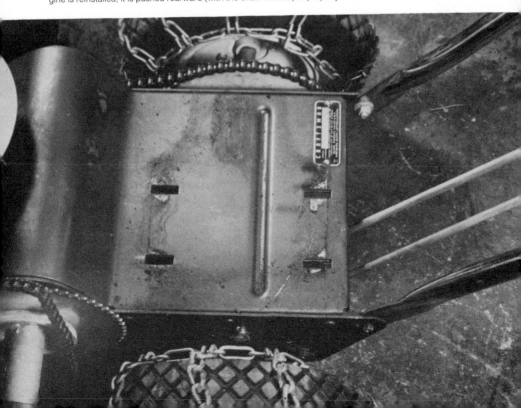

and exerts the push against the valve stem. The tappet also is called a valve lifter or cam follower.

Once these covers are off, the valves can be checked for proper clearance, then disengaged from their springs and removed. All carbon deposits can be scraped from the top of the piston, the combustion chamber and the valves. Additionally, the valves and the cylindrical parts against which they close (called seats) are resurfaced to insure that they seal off the combustion chamber completely when they are closed.

The springs that pull the valves closed are checked, and the clearance between the valve stem and the tappet again is measured and corrected if necessary. All parts are then refitted, except for the installation of a new cylinder head gasket.

Why it's done

A carbon and valve job is not an "automatic" service required after "so many" hours of operation. It's done when and if it's needed, and the reasons it may be needed are these:

1. The accumulation of carbon in the combustion chamber does not allow the combustion chamber to cool properly, with the result that some of the air-fuel mixture is ignited by combustion chamber hot spots instead of the spark plug. This hot spot ignition (called "ping" in automotive circles) is not timed, as is the spark from the plug, so the mixture ignites at the wrong time and the engine produces very little power.

2. The hot spot ignition also produces unacceptable loads on the piston's connecting rod and the crankshaft, because it occurs when they are not in the ideal positions for accepting loads. The result is premature engine wear.

3. The failure of the valves to seal the combustion chamber means that some of the air-fuel mixture leaks out during the compression stroke. The reduced compression and lost air-fuel mixture results in a less-powerful explosion when the spark arrives, and the engine loses power.

The inspections of the springs and valve clearance are extremely important. As an engine is run over a period of time, the popping closed of the valves against their seats tends to push the valve down, slightly reducing the designed-in clearance between valve stem and lifter. This clearance is necessary to compensate for the expansion of parts as the engine warms up, and if necessary clearance does not exist, the valve may seem closed, but actually may be just a little bit short of full contact with its seat. Inasmuch as the valve cools by transferring heat to its seat in the engine block (and the flywheel cools the block), the valve's failure to make complete contact reduces its ability to transfer heat. As a result, the valve burns prematurely. The burning results in erosion of the valve surface and further reduces its ability to seal properly, permitting some of the compressed air-fuel mixture to leak out.

How it's done

A valve job begins with removal of the cylinder head, so let's take a close look at the cylinder head on the typical flathead engine. As the name "flathead" implies, it's a basically flat slab of metal with a spark plug hole and a cup-like recess (the combustion chamber) in the side against the top of the cylinder. It may be held to the engine block by bolts or studs and nuts, all of which are obvious and present no problem in removal. See 5-22 through 24.

You must be careful to keep head bolts in some sort of order, for on many engines (Briggs and Stratton a major example) different length bolts are used. If you insert a short bolt in a long hole, not enough threads may catch for the kind of holding power that will survive many hours of normal engine vibration. If you try to tighten down a long bolt in a short hole, you may ruin the bolt or hole, or the bolt may go through a hole in head and block and come into contact with another part, and possibly damage it.

A simple method of noting which goes where is to draw the rough outline of the head on a piece of cardboard and little circles within the outline to represent the bolt holes. When you remove a bolt, push it through the cardboard at the appropriate circle.

Once you have them off, the cylinder head and its gasket can be lifted away from the top of the engine block. Never reuse the gasket, for even if it looks perfect, it isn't. Any gasket is used to compensate for minor irregularities in the finish of the surfaces between which it is placed. When it is bolted down, the gasket takes a set between the surfaces. If you could look at it under a microscope you would see millions of tiny depressions; from those you might find a difference in thickness of as much as .002 inch or more. Some other gaskets in a small gas engine may be reusable (although it's poor practice), but the cylinder head gasket must hold in the pressure of the compressed air-fuel mixture. If you reuse a head gasket, there is no way you could exactly realign it so that all the tiny depressions would be in exactly the same spots. As a result, compressed air-fuel mixture would leak out of the cylinder, reducing engine performance.

When you remove the cylinder head, you should check it for warpage. See 5-25. An excessively warped head will not provide a proper seal, even when tightened down with a new gasket underneath. To see if the warpage is within acceptable limits, place a straight edge, such as a good machinist's ruler, across the flat surface at several points (except the combustion chamber of course, which is not flat), and try to fit a .003-inch

5-22. A valve job on any four-stroke-cycle engine begins at the cylinder head. Removing the spark plug first isn't absolutely necessary, just customary. The engine shown is a Tecumseh.

5-23. Next, remove the cylinder head bolts, which are easily accessible. This should be done only when the engine is cold.

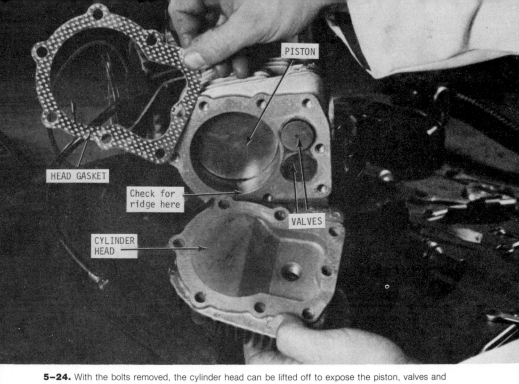

PISTON

HEAD GASKET

Check for ridge here

VALVES

CYLINDER HEAD

5—24. With the bolts removed, the cylinder head can be lifted off to expose the piston, valves and head gasket. Check for a hard-carbon ridge at the top of the cylinder.

5—25. Check the flatness of the cylinder head surface by laying a straight edge, such as a machinist's ruler, across sections of flat surface. Then try to slip a .003-inch feeler gauge underneath. If it fits underneath at any point, the cylinder head should be resurfaced at a machine shop. Repeat the procedure across the mating flat surface of the cylinder.

feeler gauge between the straight edge and the flat surface of the head. If the gauge can fit under at any point on the head, take the head to an automotive machine shop for resurfacing or replace the head. If the engine is an old one that you're just trying to wring some extra life from, recheck the warpage with an .005-inch gauge, and if the .005-inch gauge will not pass under, use two new head gaskets, one on top of the other. They should provide an adequate seal. The use of two head gaskets will increase the distance from the top of the piston to the top of the combustion chamber a bit, and the result will be that the air-fuel mixture will not be compressed into as small a space. This reduction in compression will have a small effect on engine performance, but perhaps not more than you can accept.

Before doing any major work, turn the flywheel to lower the piston to the bottom of its stroke. This will permit you to inspect the cylinder bore for scores and a ridge at the top. If the score marks are deep enough to catch a fingernail in, the engine must be disassembled, the cylinder rebored, and an oversize piston installed. This is undoubtedly a more expensive job than you might want to have done—but *have it done* you must, for very expensive special equipment is necessary. Before proceeding, read Chapter 9, "Deciding to Repair or Replace." A hard carbon ridge at the top of the cylinder is normal after many hours of operation, but it's not acceptable. See 5-24. It can be removed safely with a ridge reamer, another tool you're not likely to have. Even if you happened to have an automotive ridge reamer, it probably won't fit into the small gas engine cylinder. You should be able to find an automotive machine shop with the right size reamer to do the job. It will cost only a few dollars.

Removing the valves

On engines so equipped, disconnect the engine breather hose from the tappet cover. This hose vents internal engine fumes to the atmosphere. See 5-26. Then remove the tappet cover itself to expose the valve stems, springs and retainers. See 5-27 through 29. On the Briggs engine, also note the breather disc valve, which should be checked as shown in 5-30.

Begin by turning the flywheel until both valves are closed and the piston is at the top of the cylinder. Insert the appropriate thickness feeler gauge between the tip of the valve stem and the lifter. (Ask the parts supply store to look up the valve clearance specification for you.) Normally the clearance for intake and exhaust valves is different, but it's easy to tell which is which. Just look at the mushroom end of the valves in the combustion chamber, and whichever is slightly larger in diameter is the intake valve.

You should be able to slide the feeler gauge in and withdraw it with just light to moderate drag. It's possible the gauge fit will be loose; but the possibility is so remote, that you shouldn't even think about it. This only oc-

5-26. On the Briggs and Stratton engine there is a breather hose that must be disconnected before you remove the valve cover.

5-27. Next, remove the valve cover, held on the Tecumseh by two hex-head Phillips screws. You can use either a wrench or Phillips screwdriver.

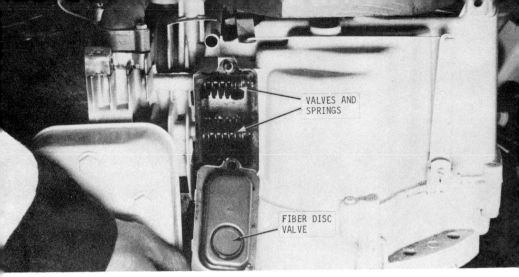

5–28. When the Briggs cover is removed you can see the valves and springs, and the fiber disc valve, part of the breather assembly.

5–29. Here Tecumseh's engine valve cover has been removed. This setup is the same as the Briggs except for the breather, which is located elsewhere. Before further disassembly of any four-stroke cycle engine, you should turn the engine by hand to see if the valves open and close completely. This is a simple check of the camshaft and valve springs.

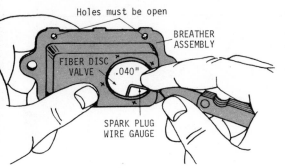

Holes must be open

BREATHER
ASSEMBLY

FIBER DISC
VALVE .040"

SPARK PLUG
WIRE GAUGE

5–30. To check the Briggs breather disc valve, try to insert a .040-inch wire gauge between the fiber disc and the inside of the cover, as shown. (This .040 inch is a common automotive spark plug size, so you can use a plug gauge if you have one.) The gauge should not fit in if the disc is functional, so don't use a good deal of force trying to get it in. The illustration is of the early type valve cover, with venting holes that must be clear. The hose shown in 5–26 is the current design.

curs if the engine had previously been rebuilt and someone made a mistake grinding the valve stem tip, or if the camshaft lobes have worn down. More likely, the valve clearance will be inadequate, and this can be corrected. Don't even think about this job yet. Wait until you're ready to reinstall the valves, for the valve clearance may be further reduced by later service, or you may decide to discard the valve.

Note: On Briggs and Stratton engines, use the following procedure to check clearance: Turn the flywheel until one of the valves moves out (opens) to the maximum position. With chalk, make reference marks on the edge of the flywheel and at the point on the engine directly across from it. Rotate the flywheel 360 degrees until the marks line up again, and check the valve clearance. Repeat the procedure for the other valve. Briggs and Stratton valve tappet clearances are listed on page 84.

Now remove the valve springs and retainers. There are three common setups for retaining the valves and springs shown in 5–31; these are 1) a pin through the valve stem, 2) a split collar, 3) a single-piece retainer with a keyhole-shaped center (the most popular for small engines).

The professional shop method is to use a valve-spring compressor, so that the coil spring can be squeezed together, permitting easy removal of the retainer, as shown in 5–32a and b. You probably don't have this tool or anything like it, and it wouldn't pay to buy one, even though it's not expensive. On an automobile engine you don't have a choice, for the valve springs are very strong. The typical small mower or blower engine, however, has reasonably light springs, and they can be pried up with a screwdriver (5–32c). With the single-piece retainer, it's possible to pry up on the

5–31. There are three common devices for retaining the valve spring to the valve. One is a simple pin through a hole in the valve stem. A second is a pair of conical collars that fit into a recess in the valve stem, held in place by the pressure of the spring; this is the design most commonly used in automobiles. The third, most popular on mowers and blowers, is the circular retainer with the keyhole and the recess in the valve stem. The retainer is inserted onto the valve stem, so that it locks in the smaller opening.

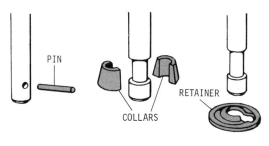

5–31a. This chart lists Briggs and Stratton Valve tappet clearances.

Model Series	ALUMINUM CYLINDER				
	Intake		Exhaust		
	Max.	Min.	Max.	Min.	
6B, 60000, 8B, 80000	.007	.005	.011	.009	
82000, 92000, 100000					
130000, 140000, 170000, 190000					

Model Series	CAST IRON CYLINDER				
	Intake		Exhaust		
	Max.	Min.	Max.	Min.	
5, 6, 8, N, 9, 14, 19	.009	.007	.016	.014	
190000, 200000					
23, 230000, 240000	.009	.007	.019	.017	
300000, 320000					

spring while simultaneously using the tip of the screwdriver to move the retainer so that the valve stem is in the enlarged section of the keyhole. Once you do this, you can release pressure on the spring and the assembly will come apart, permitting you to lift the valve up and out from the combustion chamber.

The other two designs are best tackled with a helper. One person pries up on the coil spring, and holds it up while the other releases the retainer. The pin-type is pulled with needle-nose pliers (5–33); the split collar is flicked apart with a small thin screwdriver (5–34). It should be pointed out that these two designs are somewhat easier to take apart than reassemble, so have some patience (and a helper) when the time comes.

Once you've removed the valves, retainers, and springs (5–35), reinsert

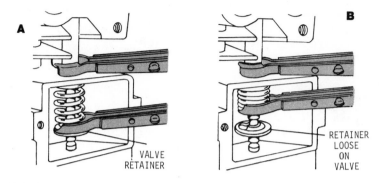

5–32. In this design, the spring is compressed either with a special tool or with a screwdriver until the retainer is loose on the valve. You then push the retainer with a screwdriver or with your finger so the larger portion of the keyhole slot is around the valve stem, permitting it to be slid down off the stem as the spring is released.

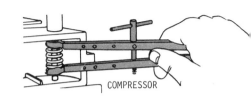

COMPRESSOR

5-33. In this illustration the spring is compressed and needle-nose pliers are inserted to pull out the pin.

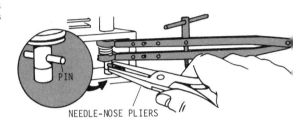

PIN

NEEDLE-NOSE PLIERS

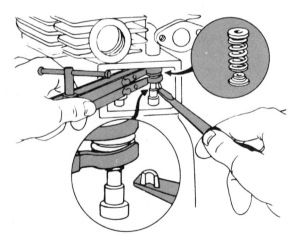

5-34. This shows how to remove the conical collar retainers. The spring is compressed and the collars are pushed apart and out with a thin screwdriver. During reinstallation, shown in the lower circular inset, coat the inside surface of each collar with grease, so it will hold in place on the stem recess as you release tension on the spring.

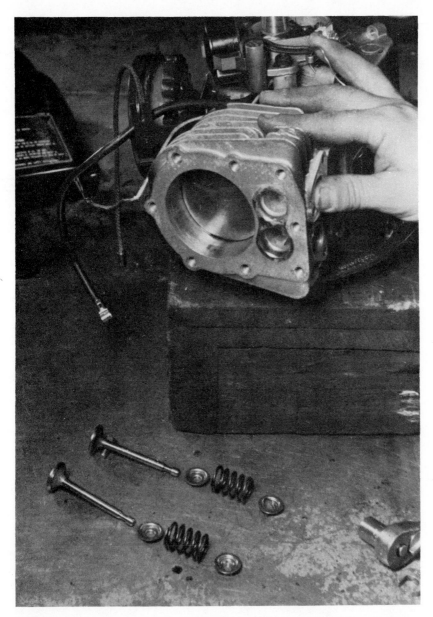

5–35. This photograph shows the parts of the valve train removed for a valve job: the valve, the upper retainer (not always used), the spring, and the lower retainer. Reinsert the stem of each valve into its guide and, holding it by its head, try to rock it sideways. If you feel any looseness, the valve guide is worn, and professional service is required. A shop will have to either replace the guide or enlarge it and install a valve with a thicker stem. The normal running clearance of a valve stem in its guide is just a few thousandths of an inch. This can be checked precisely only with special gauges.

each valve into its cylindrical guide and, holding it by the mushroom head (about a half inch up from the valve seat), attempt to move it side to side. If you feel any real movement, it's possible that the valve guide is worn. You can't just ignore this, for the valve stem that is loose in the guide will permit the engine to burn excessive amounts of oil. (The suction in the cylinder will draw oil up through the gap between the valve stem and the guide.) A machine shop may be able to enlarge the guide to permit you to use a valve with a thicker stem, if you can furnish the thicker valve. Or the valve guide may be replaceable, and this is another machine shop operation requiring a press.

Another procedure, recommended by Briggs and Stratton, is to have a service shop enlarge the guide with a reamer and then have a special insert installed. This insert accommodates a standard-size valve. In addition, the insert must be smoothed with a special finishing reamer. Owing to the special tools required, this job is definitely not one for the do-it-yourselfer.

If in doubt, ask the machine shop to check the clearance with a dial indicator, a precision instrument. The maximum reading should be .018 inch, which is equal to .009 inch (half the total) as an all-around-the-stem clearance. The clearance is most critical on the intake valve, for when this valve is open, the cylinder has a vacuum (piston dropping down on intake stroke).

If you're just trying to get a bit more life from an old engine, with minimum expense, you can just take the valves into the shop and ask for a pair of tight-fitting valve stem rubber seals. These seals may be available specif-

---

5–36. After professional resurfacing or hand lapping by you, the valve head margin and valve seat dimensions should be checked with a machinist's ruler to see if they are within the general specifications given in the illustration. These are for a Briggs engine but are applicable to any four-stroke small gas engine. The angles mentioned (45 or 30 degrees) must be checked at a machine shop.

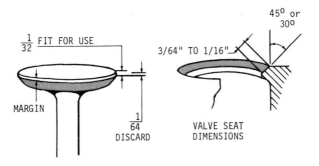

ically for your engine; or if not, it's always possible to find something that will fit. When you reinstall the valve, fit the seal (a little rubber ring) over the stem, working through the tappet chamber. Move the seal along the stem as far toward the head as possible, and then slip in the spring. Finally, move the seal still further until you can get it virtually against the valve guide.

Inspect each valve carefully (5–36), noting the condition of the valve face in particular. If it is eaten away at any part of the circumference or in the face surface that comes in contact with the valve seat, discard it. In most cases, the exhaust valve (which runs hotter) will be in far worse shape than the intake valve. Check the seat contact surface, and if it has pockmarks, it must be professionally resurfaced with special tools, as shown in 5–37 and 39. The price for such work is usually no more than about $1 per seat.

You could also have a valve resurfaced if it is pockmarked, but the difference in price between a new valve and resurfacing is so little that it hardly pays. Usually the resurfacing can be done at any automotive ma-

5–37. If the valve seat has deteriorated, you can either replace it for a few dollars or have it resurfaced for about $1. As shown in the photo, a special cutting tool is used.

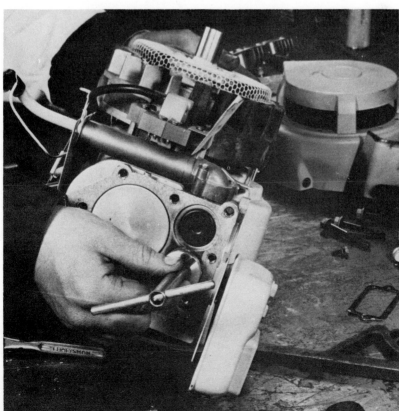

chine shop. But if a valve seat is really bad and must be replaced, the automotive shop may not have the replacement part. In any case, however, installing a new seat takes equipment you do not have and should not invest in. The total price for the job is only a few dollars.

New seat, old valve; new valve, old seat, or old and old—the valve should be mated to the seat for good sealing by a procedure called lapping. This consists of smearing a thin coating of a grinding compound on the valve face, as in 5–38, inserting the valve into the guide and down into position without spring or retainer, then oscillating the valve back and forth and around until the grinding compound wears away little bits of valve and seat that prevent full contact. The valve is then removed, the compound wiped off both parts, and the job is done.

There's a bit more to lapping than that, but not much. The compound comes in a small jar or tube and is available at most auto supply stores for a dollar or less. The lapping-in motion consists of short clockwise and counterclockwise arcs for about 15 seconds. Then you lift the valve and turn it clockwise about a half-turn to reposition it. Repeat the short clockwise and counterclockwise arcs for 10 seconds, then about another half

5–38. If the valve and seat appear to be basically sound, or if you are installing new valves, begin the lapping-in process by coating the valve head with a thin film of grinding compound, available in any auto supply store.

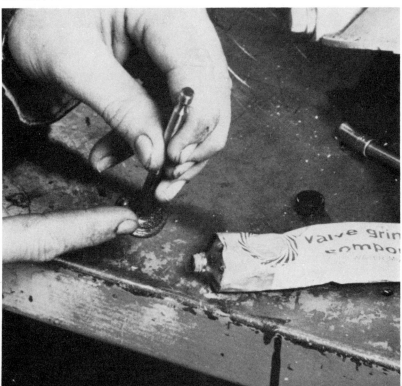

5–39. This shows the valve lapping tool with its suction cup on the valve, which has been reinserted in its guide and is now being rotated against the valve seat. Here, you see a professional valve grinding tool. But you could also use a child's toy arrow and manipulate it between your palms or insert it into a hand drill.

turn. A complete lapping operation takes about a minute or so, and in professional shops is done with the valve grinding tool shown in 5–39. It looks like a hand drill with a suction cup tip, but it provides the arcing motion. Because the suction cup holds the valve you just lift the valve off its seat with the tool and turn to reposition it.

Instead of using the special tool you can use a child's suction cup arrow or get a valve lapping tool that resembles the arrow for perhaps half a dollar. You wet the cup and push it onto the valve, the same way the special tool does it. Instead of turning a crank, however, you simply rotate the arrow stem between your palms. Maintain slight pressure on the valve so that the lapping compound can grind away at the valve and seat.

After wiping the valve and seat with a rag to remove the grinding compound, check your work. A simple check to insure that the valve and seat are really sealing is to draw lines across the width of the valve face with a soft pencil—about one line every eighth of an inch around the circumference of the face. Insert the valve, attach the lapping tool or suction arrow, apply some downward pressure and rotate the valve through a couple

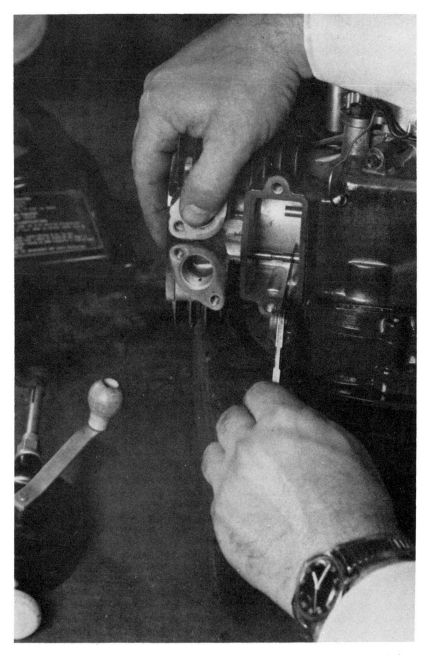

5–40. Check the valve clearance with a feeler gauge prior to reassembling the parts. Intake and exhaust valves operate with different clearances, so check specifications at your parts supplier.

5–41. If the valve clearance is inadequate, grind the end of the valve as shown. Grind for just a few seconds at a time, then recheck it by holding the valve down with thumb pressure on the flat head of the mushroom section.

of arcs, as if you were beginning a lapping operation. Remove the valve and look at the face. All the pencil marks should be just partly erased. If any marks are not, recoat the valve face with compound and go through a complete lapping operation.

Insert each valve into position. Then hold down and check the clearance with a feeler gauge once more. See 5–40. If the clearance is inadequate, you must grind the end of the stem to increase it. This is normally done on a shop grinder (5–41), but if you don't have this piece of equipment, you can get by with a fine-grit grinding wheel bit from your electric drill. Mount the drill firmly, preferably in a vise.

Cleaning

Depending on operating conditions, there may be a fair bit of carbon and oil deposit to scrape off various parts. The major accumulators are the top of the piston, the valve face and seat, the joint between the valve stem and the mushroom head, the valve springs, and the combustion chamber. Valve deposits are most easily cleaned with a wire brush, either in a shop grinder or on an electric drill. Other parts can be scraped with a small screwdriver, but be careful not to score them. Chip off any chunks with the screw-

Spring must be square

5–42. To check the valve spring, stand it up and place a carpenter's right angle level against it. If the spring is square, the edge of the level will touch all coils of the spring. If coils at the top or bottom do not touch while others do, replace the spring.

FREE
LENGTH

driver, then do the final cleaning with a small wire brush—on the drill if you're careful, by hand if you're worried.

Valve Springs

The tension of valve springs can be measured on a special fixture but even some very good shops may not have it. And even if they do, the factory specifications might not be obtainable. Such specifications are readily available for cars which have 12 to 16 springs, but not for small gas engines which have just two. Apparently the theory is that if the valve springs feel weak, replace them.

Your only possible check of a spring, therefore, is for squareness (5–42). Line it up against the vertical side of a right angle such as a carpentry level, and if all the coils touch the side of the level, the spring isn't cocked. Then press down on the spring, and if it doesn't feel weak, reinstall it. One way to get a standard of comparison is to buy a new set of springs and press down on the new, then the old. If you can't feel any real difference, return the new spring. Most parts suppliers will take them back for, at most, a 10 percent handling charge.

Note: Some engines have different springs for intake and exhaust valves. To avoid a mixup, look at each spring as you remove it, and if one looks different from the other (number of coils, length, etc.), make sure it goes back on the same valve from which it was removed.

REASSEMBLY

Although you may struggle a bit with the valve springs, once they're right, you'll know it. The part of the job that takes the most "touch" is tightening the cylinder head nuts or bolts.

All small gas engine manufacturers specify both a tightening sequence and a specific amount of tightening, given in pounds feet or in pounds inches (divide by 12 to obtain pounds feet if your torque wrench is so calibrated) and measure with a torque wrench. Using a torque wrench helps you insure the even tightening that means a properly compressed cylinder head gasket, and therefore full compression.

Drawings 5–43 to 45 show typical Briggs and Tecumseh tightening patterns. If you don't have a tightening pattern, start at the center and move outward, first a bolt on the left, then one on the right, then the next farthest on the left, next farthest on the right, and so on.

Whatever the tightening specification, divide it by three and tighten all bolts or nuts in approximately equal stages. See 5–43 through 46. Example: The specification is 165 pounds inch, equal to 14 pounds feet. Tighten all nuts or bolts to five, then to 10 and finally 14. This helps to insure even seating of the cylinder head gasket.

If you don't have a torque wrench, don't want to buy one and can't borrow one, all you can do is try to tighten in stages by feel, making sure that when you finish all the bolts are tight without your "leaning" on the wrench. I can't say that I like this procedure and would recommend that if all else fails you try a tool rental store, where the charge for a torque wrench for a day might be only a couple of dollars.

VALVE JOB (2-STROKE-CYCLE ENGINE)

On a two-stroke-cycle engine the valve is a reed, which you should replace if it is loose, or if its screw holes are elongated, or if its contact surface is

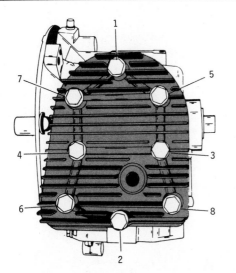

HEAD BOLT TIGHTENING

5–43. This is the head bolt tightening sequence for all Tecumseh flathead engines except eight-horsepower models. The bolts should be tightened in stages, first to 5 pounds foot, then to 10 and finally up to 12 to 16.

HEAD BOLT TIGHTENING (Continued)

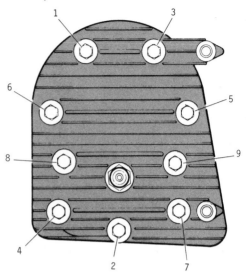

5-44. In the head bolt tightening sequence for the eight-horsepower Tecumseh flathead engine, tighten all bolts to 16 to 17 pounds foot in three stages (first to 5, then to 10, finally to 16-17 pounds foot).

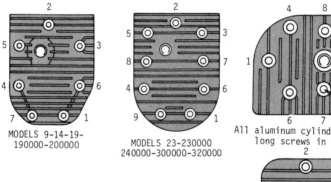

MODELS 9-14-19-190000-200000

MODELS 23-230000 240000-300000-320000

All aluminum cylinder engines have long screws in these 3 holes

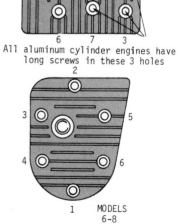

MODELS 6-8

5-45. These are the head bolt tightening patterns recommended for Briggs and Stratton engines.

ALUMINUM CYLINDER	
Model series	Lbs. ft. torque
6B, 60000, 8B, 80000, 82000, 92000, 100000, 130000	11 to 12
140000, 170000, 190000.	13 to 14
CAST-IRON CYLINDER	
Model Series	Lbs. ft. torque
5, 6, N, 8, 9	11 to 12
14	13 to 14
19, 190000, 200000, 23, 230000, 240000, 300000, 320000	16 to 17

5–46. This chart provides Briggs and Stratton head bolt torque specifications.

damaged; see 5–119. If the sealing surface against which the reed bears is gouged or scored, replace it. If the sealing surface and reed have merely accumulated dirt, clean them with automotive solvent. Before you reinstall the reed, coat reed screw threads with Loctite.

CENTRIFUGAL CLUTCH JOB

A clutch replacement on an automobile is a major job, but the centrifugal clutch on the typical small gas engine (except simple rotary mowers) is relatively easy. See 5–47. You'll find this clutch on the end of the crankshaft opposite the flywheel end, and in most cases you can get to it without major disassembly. This is particularly true on chain saws, for the clutch is on the side of the chain, to which the manufacturer invariably provides easy access.

There are many kinds of clutch mountings. Some are held in obvious ways (a nut or bolt), others in ways not so obvious, such as the hidden Allen-screw type shown in 5–88.

Removing the clutch's threaded retainer may present some minor problems, and in some cases, such as shown in 5–107, a spanner wrench is necessary. It's a tool you probably don't have. I know of only one possible substitute for the spanner, namely teaming up a pair of needle-nose pliers and vise-type locking pliers. If the needle-nose tool is of an adequate size, you can spread the jaws to fit into the holes, then lock the vise pliers on the needle-nose pivot. Use the handle of the vise-type grips as a lever arm. If you don't have the needle-nose pliers and want to get the spanner, it won't be a budget-buster. However, the spanner is not a tool you'll get a lot of use from, so don't buy until you need it.

CENTRIFUGAL CLUTCH

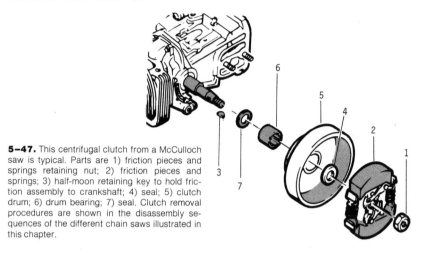

5–47. This centrifugal clutch from a McCulloch saw is typical. Parts are 1) friction pieces and springs retaining nut; 2) friction pieces and springs; 3) half-moon retaining key to hold friction assembly to crankshaft; 4) seal; 5) clutch drum; 6) drum bearing; 7) seal. Clutch removal procedures are shown in the disassembly sequences of the different chain saws illustrated in this chapter.

As with flywheel removal, a good procedure when taking out a clutch is first to remove the spark plug. Next lower the piston by turning the flywheel; then stuff the cylinder area on top of the piston with clothesline to serve as a shock absorber. Clutch disassembly sequences are shown in 5–84 through 88, 5–105 through 110, and 5–147 through 151.

What the clutches all have in common is some type of coil or leaf spring arrangement holding pieces of friction material in a hub bolted to the flywheel. This assembly is covered by a drum on a bearing that rides on the crankshaft. This bearing may be a roller type, as shown in 5–47, or perhaps a plain bearing. When engine speed reaches a designed-in figure, centrifugal force overcomes the springs, and the friction material parts are forced outward against the inside circumference of the clutch drum, locking the drum to the flywheel. The drum may have a hub to which a mower blade is bolted, or a sprocket around which a chain is wrapped.

When disassembling a clutch with chain and sprocket, check the condition of the sprocket teeth. If they are badly chipped, the chain soon will slip and possibly snap or damage the mechanism it drives. If the drum is a two-piece design with a cover, you can take it off and inspect space between the drum surface and the friction material. That space should be uniform all around. If it isn't, the clutch action has not been consistent. The probable causes are a weak friction material spring, a scored drum surface or a defective drum bearing.

Spin the drum by hand and listen for any noise from the bearing. When

MUFFLER REPLACEMENT

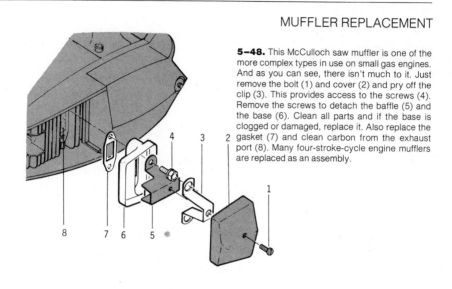

5–48. This McCulloch saw muffler is one of the more complex types in use on small gas engines. And as you can see, there isn't much to it. Just remove the bolt (1) and cover (2) and pry off the clip (3). This provides access to the screws (4). Remove the screws to detach the baffle (5) and the base (6). Clean all parts and if the base is clogged or damaged, replace it. Also replace the gasket (7) and clean carbon from the exhaust port (8). Many four-stroke-cycle engine mufflers are replaced as an assembly.

BRIGGS AND STRATTON ENGINE

5–49. This is the Briggs and Stratton four-stroke engine with some external parts labeled.

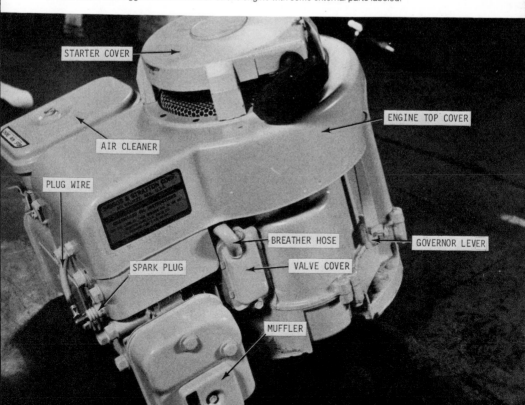

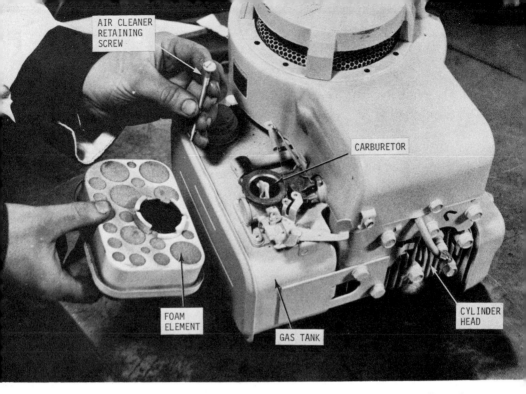

AIR CLEANER
RETAINING
SCREW

CARBURETOR

FOAM
ELEMENT

GAS TANK

CYLINDER
HEAD

5-50. Disassembly of the Briggs and Stratton engine begins with removal of the air filter, a foam element in the canister, held to the carburetor by a long screw.

5-51. Before doing major engine work or draining oil, remove the engine oil drain plug by inserting a screwdriver between the two tangs of the plug, as shown. Once the plug is out, tilt the engine to let the oil drain from the plug hole into a pan.

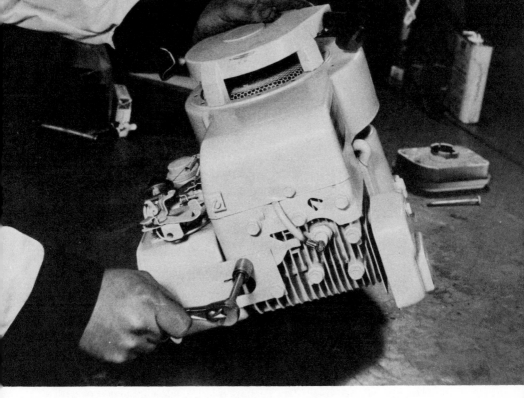

5-52. Fuel system service, if it includes gas tank removal, begins with unthreading the gas tank bracket bolts on each side of the engine. This job is also necessary if you plan to remove the cylinder head, because the tank bracket bolts on the cylinder head side of engine also serve as head bolts.

you remove the drum, inspect the bearing, looking for a blue tinge which would indicate heat damage. If the bearing is the popular roller or ball type, hold the inner cage by the edge and spin the outer, to cause the balls or rollers to move. If you feel any irregularity in movement, or a blue look, replace the part. Otherwise, pack with lightweight automotive grease (such as Lubri-Plate) and reuse.

If the clutch drum is scored on the surface that comes in contact with the friction material, replace it. To make the distinction between light scratches (which can be smoothed by sanding with a mild abrasive called crocus cloth) and scores, try to catch a fingernail on the marks. If the fingernail catches, replace the part.

The availability of detail parts determines what you do with worn friction material or apparently weak springs. More likely than not, you'll have trouble getting anything but a friction material and spring assembly, so that's what you'll use. There's good logic for replacing the friction material and springs under any adverse conditions. If the clutch has had a long and useful life and the friction material is now worn out, odds are that the spring has also weakened with age. If you don't change it, you'll get relatively low speed clutch engagement, making it difficult to adjust the idle speed on a chain saw, for example. If the clutch has failed prematurely, the friction material is going to be replaced. So you might as well not take chances by trying to reuse the spring, which might be responsible for or at least a contributor to the failure.

Centrifugal clutches, it should be pointed out, are normally very reliable parts that can last the life of the engine. Unless your troubleshooting points to a clutch problem and you see something wrong (worn friction material, scored drum, clutch grabbing at low speed, or failure to release when the engine stops), don't blame the clutch for much.

MUFFLER REPLACEMENT

When replacing an automobile muffler, you have to jack up the car, then hammer and chisel away to remove the muffler. But the small gas.engine is a snap for muffler replacement. Some mufflers are removed as easily as unthreading a screw or two. The most complex have a cover held by two screws, then a main body held by two more which are accessible after the cover comes off. See 5–48. The Tecumseh four-cycle engine comes with a muffler held by screws, but the exhaust port is threaded so that the screw-in one-piece muffler used on some Briggs and Stratton engines also can be installed. This is a handy thing to remember if one of the Tecumseh muffler screws freezes in place and snaps off when you apply wrench pressure. You don't have to drill out the hole, tap and insert a thread repair coil; just use the thread-in muffler as shown in 5–73a. When removing the muffler on a chain saw in particular, clean carbon from the exhaust port, as shown later in this chapter (5–153). A plastic stick is recommended, but you can use an ordinary screwdriver if you're careful not to dig in, which could score the port. Also take care that the screwdriver doesn't slip and scratch the piston.

ENGINE DISASSEMBLY—THE BRIGGS AND STRATTON

The Briggs and Stratton engine is the most widely-used four-stroke-cycle small gas engine. It's the engine you'll find on perhaps 70 percent of the lawn mowers and snow blowers in the homeowner model category. The smaller sizes sold to assemblers of homeowner equipment are all of the flathead type, which means that the cylinder head is a slab that holds only the spark plug. The valves that admit the air-fuel mixture and allow the exhaust gas to be expelled are in the engine block.

The engine shown in 5–49 is a vertical crankshaft type you'll find on

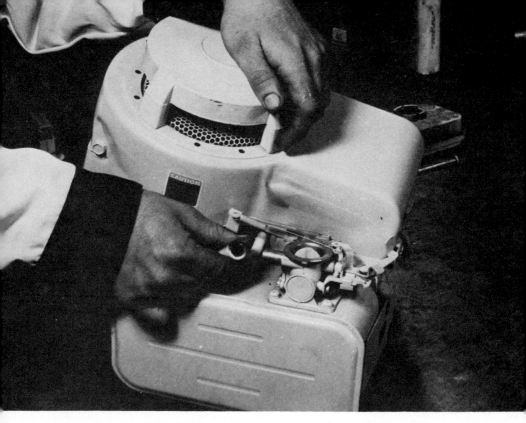

5–53. Next, disconnect the fuel line at the carburetor.

many rotary mowers. The air filter, 5–50, is a cleanable foam element, the service for which is described in Chapter 7.

If you only wish to remove the gas tank for thorough cleaning, for example, limit yourself to those steps in the sequence that apply, namely 5–52, removing the tank mounting bolts; 5–53, disconnecting the fuel line; 5–54 and 55, removing the top cover; 5–56, disconnecting throttle linkage springs; 5–57, and disconnecting the magneto kill wire. To remove the carburetor from the tank, pick up the disassembly sequence for fuel systems in Chapter 7.

Service of ignition points, condenser and magneto do not require removal of the gas tank, so simply begin at 5–55, removal of the top cover. Then proceed to 5–60, removing the gravel guard from the flywheel; then take up 5–61 through 68. Actual service procedures for the ignition system plus coverage of Briggs engines other than the one disassembled are in Chapter 6.

5-54 and 55. Remove the engine top cover bolts, left. If the cover doesn't seem to lift off easily, check to see if you've missed a bolt. Next, lift off the engine top cover, exposing the flywheel, magneto coil and other parts, including the carburetor-to-governor linkage bracket springs that must be disconnected prior to removal of the gas tank.

5-56. Here's a close-up view of key parts, springs and bracket. Before disconnecting the springs from the bracket, make a note of which spring goes into which hole.

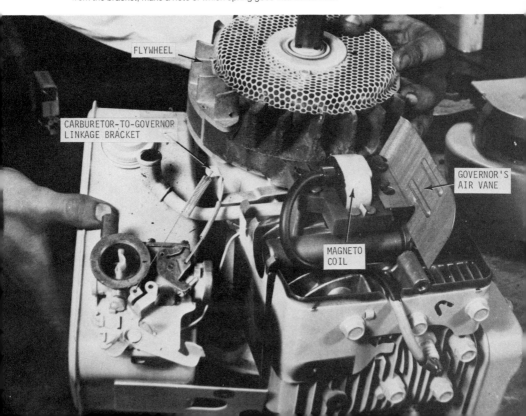

FLYWHEEL

CARBURETOR-TO-GOVERNOR
LINKAGE BRACKET

GOVERNOR'S
AIR VANE

MAGNETO
COIL

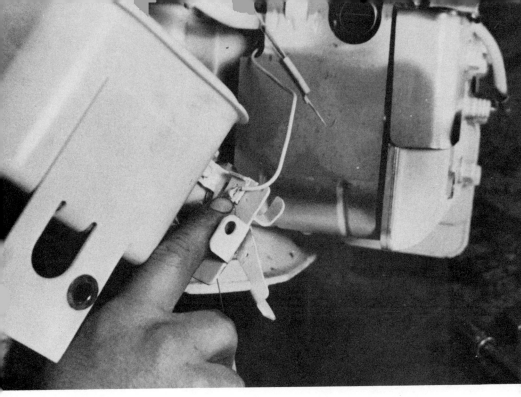

5—57. Now disconnect this wire from the tank. You might first, however, wish to note the operation of the little tab arrangement, for this is the magneto kill switch setup that stops the engine. When the throttle is closed completely, the tab moves to ground the wire to the gas tank. The wire is connected to the magneto coil, so that when it's grounded, the magneto is short-circuited, killing the spark that fires the plug. The gas tank and carburetor can be lifted away. Separating and servicing them is covered in Chapter 7.

5—58. Now you're ready to remove the remaining cylinder head bolts, and the sheet metal cover over some of them also will come off. Note that the spark plug wire has been disconnected, but the plug is still in place.

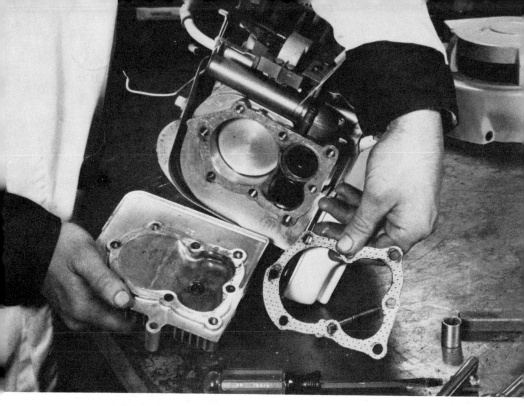

5–59. The cylinder head is off. And you see the head gasket, piston and valves. If necessary, a valve job could be performed easily at this point.

FLYWHEEL REMOVAL

5–60. Now to the flywheel. Remove the two bolts holding the gravel shield.

5—61. With the shield off, you get a clear view of the starter clutch assembly.

5—62. This is the proper way to remove the starter clutch, with one spanner tool to hold the flywheel by two fins, and a specially-shaped wrench to unscrew the clutch assembly.

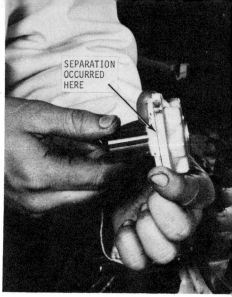

SEPARATION OCCURRED HERE

5–63 and 64. If you don't have the special tools, a hammer and brass drift whacked against the ears will loosen the clutch. You should have a helper hold the edge of the flywheel with his hands. Use the hammer and drift carefully, for very sharp whacks may distort the ears, making it impossible to insert the screws that hold the gravel shield. The right-hand photo shows the starter clutch removed. Watch for any looseness at the joint, which indicates the assembly will come apart, making necessary some reassembly.

5–65. Reassembly of the clutch begins with coating the balls inside with grease and then stuffing them into the tips of the star, as shown. The grease will hold them in place. Refit the other half of the clutch, and crimp the circumference of the housing over with the hammer and brass drift.

GREASE-COATED BALL

5–66. With the starter clutch off, the flywheel can be pried up with two screwdrivers as shown. Not all models have this arrangement, as illustrations on ignition systems show in Chapter 6.

5–67 and 68. Before actually lifting the flywheel off, check for free play by holding the crankshaft and trying to turn the flywheel clockwise and counterclockwise. The flywheel should not move. If there is any movement, inspect the half-moon key that positions the flywheel on the crankshaft, and replace the key if necessary. If the wear is in the crankshaft or flywheel, these parts must be replaced. Normally the key and the flywheel take the punishment, and these parts are easily replaced; although the flywheel is not cheap. The right-hand photo shows the flywheel off. The ignition points will be accessible after the cover screws are removed.

Although not shown in the preceding photographs of his particular engine, the clutch is on the end of the crankshaft opposite to the flywheel. So simply turn the engine over, remove the bottom cover and go directly to the clutch. Typical service of the clutch is shown in other disassembly sequences in this chapter.

A valve job does require removal of the gas tank and the engine's top cover to provide access to the side cover over the cylinder head (5–52) and the cylinder head itself. Therefore, begin a valve job with the gas tank removal sequence and then proceed to 5–58 and 59, removing the head bolts and the head. Notice that it is not necessary to remove the spark plug, although removing the plug wire prior to lifting off the head may make the job more convenient.

TECUMSEH FOUR-CYCLE ENGINE DISASSEMBLY

The Tecumseh four-cycle flathead engine disassembly is shown in 5–69 through 88. This is the engine used on Sears mowers and blowers. You should note the following points:

1. When the gas tank is unbolted from the engine, just pull it back and it disconnects by itself from the fuel line. See 5–72. When reassembling, be sure that the round neck on the tank engages the fuel line.

2. The air filter is a pleated paper element attached to the carburetor air horn by a hose. There's no way you can miss finding the carburetor.

3. You can reach up and disconnect the terminal from the magneto stop ("kill") switch, as shown in 5–76.

4. The flywheel has a gear around the circumference that meshes with a gear on the starter, which is at a right angle to it. Note: Other Tecumseh engines do not have this gear arrangement. See Chapter 6.

5. Flywheel nut removal is best done with the hand impact wrench as shown in 5–79.

5–69. This is the Tecumseh four-stroke-cycle engine used on a Sears mower.

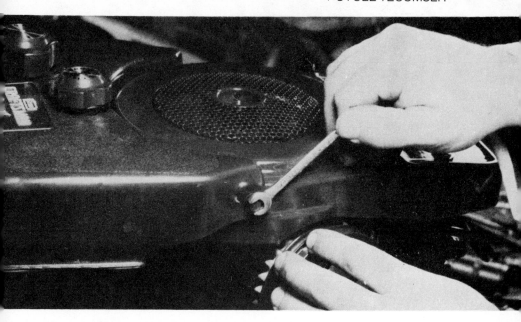

5-70. Engine disassembly begins with removal of the gas tank bolts.

5-71. Remove the pleated paper-element air cleaner from its retaining clip and swing aside the connecting hose to the carburetor.

FUEL LINE

5-72. Once the bolts are out, the gas tank can be pulled away, disengaging the tank from the fuel line to the carburetor.

5-73. Locate the bolts that hold the engine top cover. Note, they're combination hexagonal and Phillips-screw head, permitting you to use whichever tool fits in and is adequate for the job of loosening. Note the muffler, which is readily accessible without your taking off anything else.

MUFFLER

Damaged
threads
here need
not be
repaired

SPIN-IN
MUFFLER

5-73a. If the muffler retaining screws are stripped, this special replacement muffler in the photo at right can be threaded into the exhaust port.

5-74 and 75. Not all top cover bolts are obvious. This one at left is on a flange that extends to the bottom of the engine. The top cover bolt at right also serves as a cylinder head bolt.

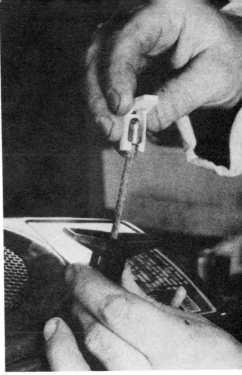

5-76. The magneto kill switch (labeled the "stop" and "run" switch) is disconnected by pulling off the spade-type connector.

5-77. Next disconnect the starter handle by separating the handle halves and removing the rope end from the top half.

5-78. The engine top cover comes off, exposing the flywheel with its gravel shield. Note that this flywheel has teeth around its circumference that mesh with teeth on the starter, which is mounted at a right angle to the flywheel.

FLYWHEEL TEETH

STARTER TEETH

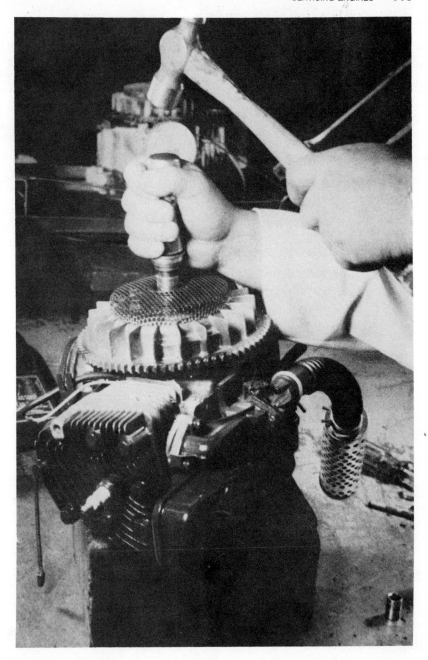

5—79. Professionals use an air-operated impact wrench, but this hammer-actuated impact wrench is a lot less expensive, does the job of removing the flywheel nut very nicely, and also serves a variety of purposes around the home.

5–80. The gravel shield is off along with the flywheel nut, but the job of removing the flywheel isn't complete. For the flywheel still is tight on a tapered end of the crankshaft. Another step is required to break the flywheel's tapered fit to the crank.

5–81. The flywheel loosening procedure consists of installing a soft-tip, closed-end nut on the flywheel, turning it down finger tight, then whacking the closed end with a hammer while you pry up with a screwdriver. Move the screwdriver around the circumference of the flywheel as you whack with the hammer. The closed-end nut is a popular "special tool" available from most small gas engine parts suppliers at low cost, but you can also do the job with two nuts (one in addition to the flywheel nut). Just make sure the second nut projects slightly above the tip of the crankshaft, so that when you use the hammer, the nut absorbs the blow.

5–82. The flywheel comes off, exposing the condenser and the ignition points cover.

5–83. Remove the cover and you have full access to the breaker points.

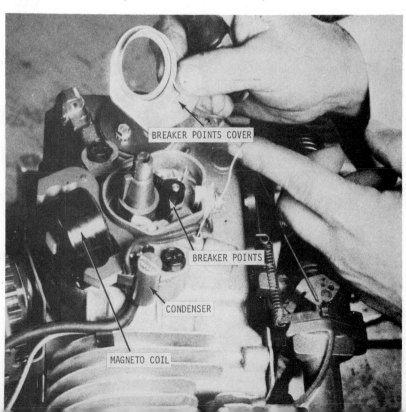

BREAKER POINTS COVER

BREAKER POINTS

CONDENSER

MAGNETO COIL

5–84 and 85. This centrifugal clutch is from a Tecumseh-powered snow blower. Disassembly begins with removal of the spring clip with a screwdriver, shown left. Then the drum comes off, exposing the hub and its friction material and the coil spring that holds it to pulley-like hub.

5–86. To replace friction material pieces, pry off the spring with a screwdriver.

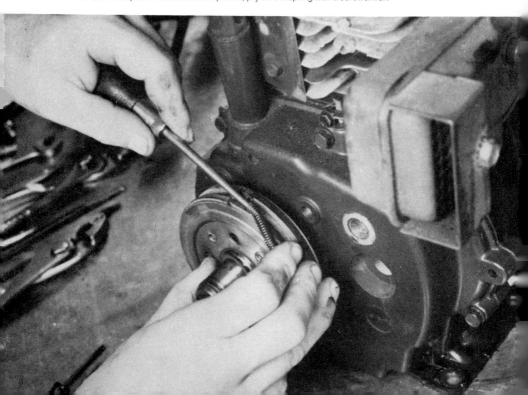

5–87. Four friction material pieces are removed. Normal replacement procedure is to install new pieces with a new spring.

5–88. If it is necessary to replace the pulley-like hub, remove the Allen-type retaining screw in back.

ALLEN SCREW

TECUMSEH TWO-CYCLE ENGINE DISASSEMBLY

This is one of the engines the pros tried to completely overhaul without the right tools and, as mentioned in the first paragraph of this chapter, it didn't work. The reason we started complete disassembly on this one was the fact that it's extremely easy to tear the carburetor gasket during carburetor removal, and the only way to replace the gasket is to unbolt the halves of the two-piece engine block. To avoid the headaches that can follow, do not replace the broken gasket; just fill in at the broken places with a nonhardening sealer, such as Permatex No. 2, and also apply a thin coat of the sealer to the exposed gasket surface. You can bolt the carburetor back on and rest assured the seal will be okay. If the gasket is badly damaged, just trim it at the engine block, trim a new gasket to fit, and you can leave the engine block halves together.

Although the engine disassembly in illustrations 5–89 through 127 is quite thorough and self-explanatory, please take note of 5–107, clutch removal with the spanner wrench. There the hammer is whacking the end of the wrench in a clockwise direction because the threads on this particular clutch are reversed (called left-hand threads). It's something you should always be on the watch for on a clutch or flywheel.

5–89. This small Sears chain saw is powered by a two-cycle Tecumseh engine. A basic maintenance job, replacement of the air filter, takes just seconds. Remove the cover and there it is.

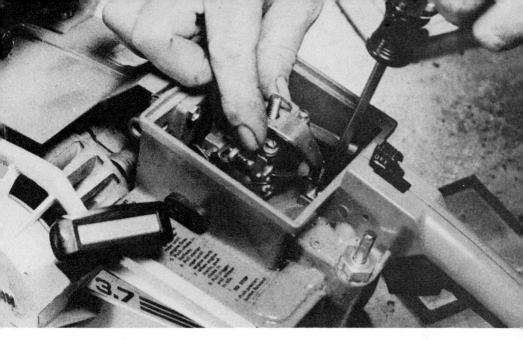

5–90. It is necessary to remove the carburetor before removing the engine. Begin by unthreading the air-cleaner-bracket screw and removing the bracket.

5–91. Next, using needle-nose pliers, you should remove the cotter pin that joins the choke knob to the choke plate link.

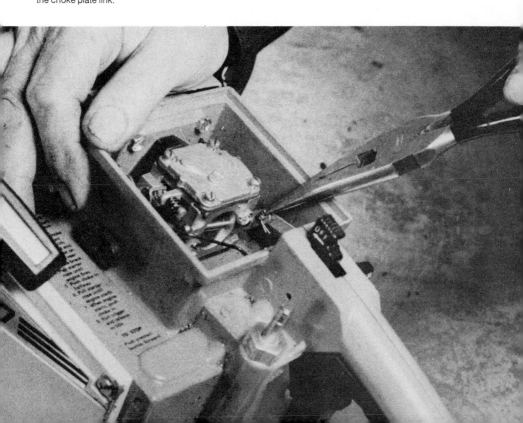

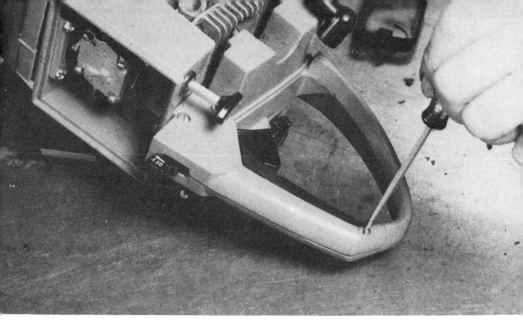

5–92. Remove the two screws that hold the cover, under which is the throttle trigger linkage.

5–93. With the cover off, the trigger linkage is now accessible.

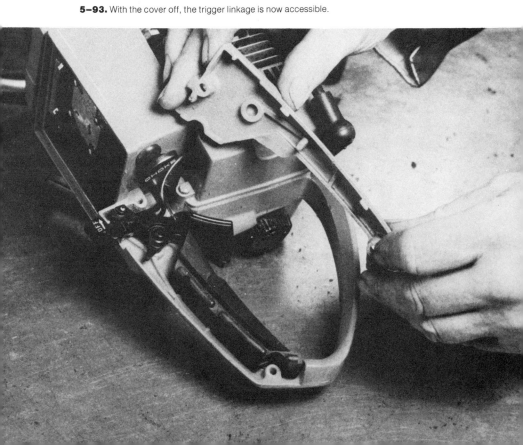

5-94. Pull the link from the throttle trigger.

5-95. Now it's a simple matter to remove the two retaining screws that hold the carburetor, using either a hex wrench, as shown, or a ratchet screwdriver.

CARBURETOR
RETAINING
SCREWS

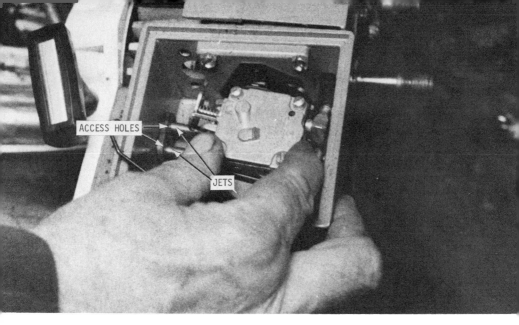

5–96. Move the carburetor sideways a bit so that the jets clear the access holes in the side of the housing; then lift the carburetor up and out.

5–97. Now remove the starter (and flywheel) side cover screws.

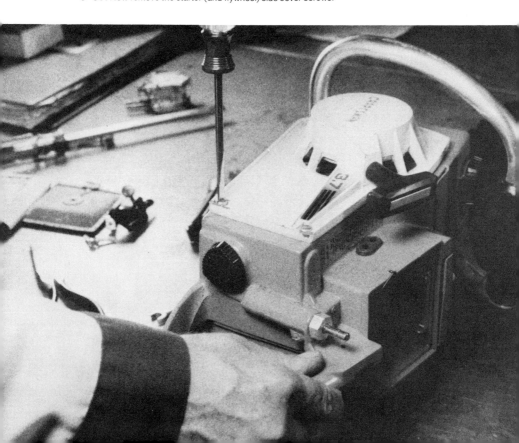

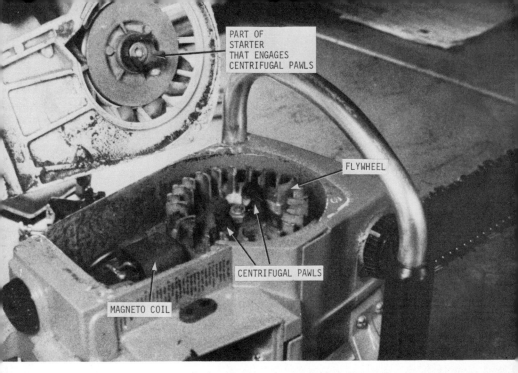

5-98. Lift out the side cover (with starter) to expose the flywheel and magneto coil. Note the centrifugal pawls in the center of the flywheel. They engage the part projecting from the starter.

5-99. If you wish to remove the engine, continue by reaching in as shown with needle-nose pliers to pull the wire terminal from the magneto kill (stop) switch.

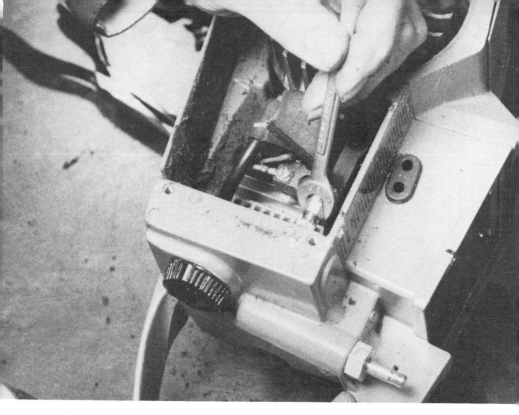

5—100. Disconnect the oiler line fitting.

5—101. Remove the muffler, beginning by unthreading the cover screws. This job can be done as a single operation if necessary to replace a defective muffler.

5-102. Here the muffler cover and insert are off, exposing the screws that hold the housing to the engine's exhaust port. Remove the two housing screws to complete the job.

5-103. To remove the chain, begin by unthreading the two chain nuts, which also hold the side cover. If you were merely replacing the clutch, you would begin at this point.

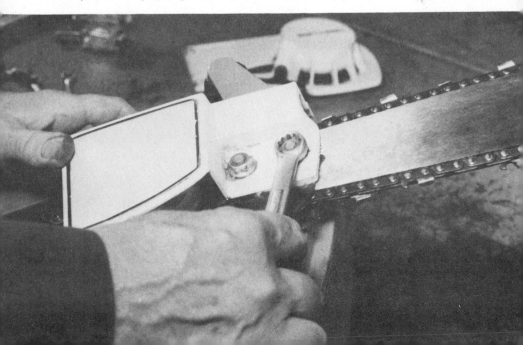

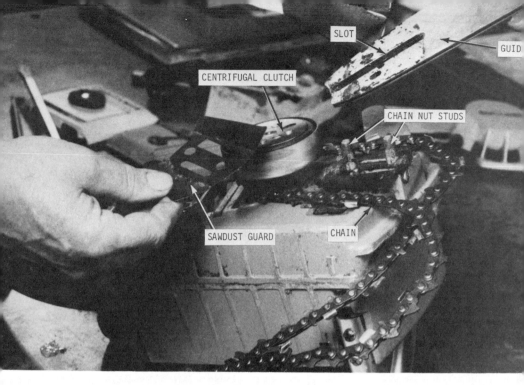

5-104. With the chain cover removed, you can see the key parts. The guide bar slot fits over studs, so that when you turn the chain-tension adjusting screw, all you are doing is moving the guide bar in or out, along its slot.

5-105. Remove the spark plug, and then go to the clutch.

5-106. Pack as much clothesline into the cylinder as possible. This clothesline serves as a shock absorber for the piston.

5-107. Attach a spanner wrench through two holes in the clutch hub. Then whack the end of the wrench with a hammer. The crankshaft will turn a bit, but the clothesline prevents damage to the piston. Note that on this engine the clutch has a left-hand thread (clockwise to loosen).

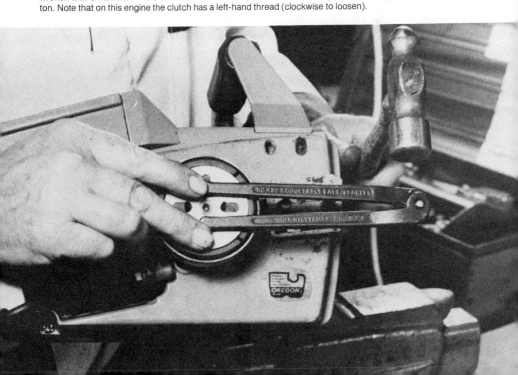

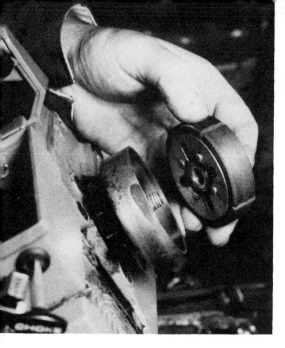

5—108 and 109. Once the friction material hub is loose, unthread by hand and remove. Then inspect the clutch friction material in profile view before removing. Look for signs of irregular wear. In the right-hand photo, the clutch friction material hub and the drum have been removed for illustration, but the actual check should be made with the clutch in place. Turn the crankshaft and watch the hub for any eccentric movement or irregular gap between friction material and edge of drum.

130

5—110. The drum pulls off, carrying its roller bearing with it. (In some cases the bearing will remain on the end of the crankshaft.) Before pulling the drum, spin it to see if it turns smoothly and easily. After removal, grease bearing rollers with a high-temperature grease, such as Lubriplate. Check the sprocket teeth for deep gouges worn in by the cutting chain, and replace the drum-sprocket assembly if necessary.

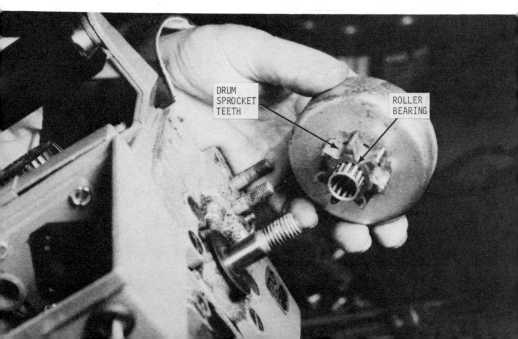

DRUM SPROCKET TEETH

ROLLER BEARING

5-111. Disconnect a second oiler line fitting, this one under the clutch.

5-112. Remove the screws that hold the flywheel-side upper-cover plate.

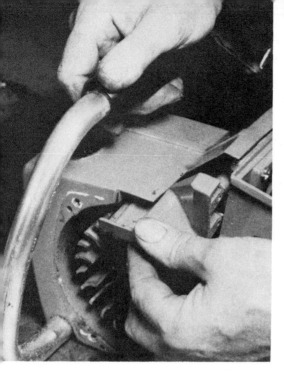

5–113 and 114. Remove the flywheel side upper-cover plate, left. Then take out the four screws that hold the rear part of the housing to the engine.

5–115. Remove the engine mounting screws. These slot-head screws are very tight, so they must be loosened (and tightened in reinstallation) with the hand impact tool, fitted with a screwdriver bit.

5–116. Take out the screw that holds the rear part of the housing to the bottom of the housing.

5–117. Remove the rear part of the housing. As you do this you will almost surely break the gasket that seals the reed plate-to-housing area, but don't worry.

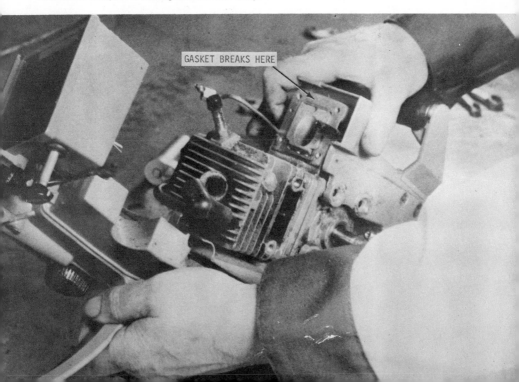

GASKET BREAKS HERE

5-118. The engine is now out of the main section of the housing and is ready for bench service.

5-119. Pull out the reed plate for easy replacement of the four reeds. If only this job is to be done, only the rear part of the housing need be removed. Reeds are held by two screws each. In this photo· you have a closer look at the surface with the gasket torn away.

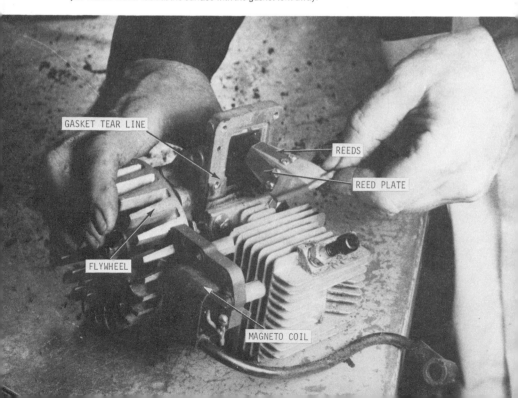

GASKET TEAR LINE

REEDS

REED PLATE

FLYWHEEL

MAGNETO COIL

5-120. The torn gasket is part of a single gasket that also seals the engine halves, so let's see what happens if you split the engine by taking off the four nuts that hold it together.

5-121. Here's the engine split apart with key parts labeled. You can see how the gasket tore away.

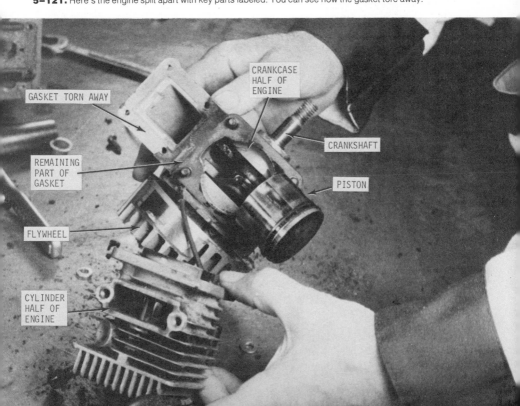

5-122 and 123. After splitting the engine, you have the problem of getting it back together, and we didn't even go as far as pulling the crankshaft and bearings. Trying to press the piston rings back together is a sure way to break fingernails and probably piston rings too. For this job, you really need a piston rings compressor. The right-hand photo shows another little item you have to look for: a locating pin in the piston. Unless the piston ring opening (gap) is at that pin, you'll surely break a ring during installation, even if you use a compressor.

5-124. This is the rear of the rear housing, where the remaining part of the gasket is. If you can, gently peel the gasket off and reinstall it on the engine. Or get a complete, new gasket and cut it at the crankcase halves joint; discard the lower half, and just use the upper portion. A third option is to cut away the old gasket at the crankcase halves line and smear the gasket surface with a nonhardening sealer such as Permatex No. 2

FLYWHEEL NUT

FLYWHEEL STARTER PAWLS

5–125. You don't have to pull the engine out for ignition service. You just remove the flywheel side cover, but this view (with the engine out) gives you a clearer picture of what to do. Use the hand impact tool shown to loosen the nut in the center of the flywheel, but do not remove the nut completely.

5–126. Whack the nut with a hammer to break the flywheel loose from its tapered fit on the crankshaft. Then pry it up and off with a screwdriver.

UNDERSIDE
OF FLYWHEEL

BREAKER
POINTS COVER

5-127. Here, the flywheel is off, and the screwdriver points to the half-moon key. Check the key's fit, both in the crankshaft and the flywheel. Two screws hold the breaker points cover, and once that's off you can service the points.

BEIARD-POULAN TWO-CYCLE ENGINE

The Beiard-Poulan two-cycle chain-saw engine, used on medium-size Sears products, is an example of an engine that is part of the exterior housing. Some service points about this engine (5–128 through 159) worth particular note are these:

1) The air filter (5–129) should only be cleaned in solvent, then blown dry with compressed air.

2) As on the Tecumseh, the clutch hub has a left-hand thread.

3) The flywheel is tightened to only 13 to 15 pounds foot of torque, so you should be able to hold it by hand when unthreading the retaining nut (5–137).

5–128. This Sears Explorer II chain saw has the Beiard-Poulan engine, which is part of the saw housing. Although the engine therefore cannot be completely separated from the housing, you can do considerable stripping of some parts of the housing and individual components.

5–129. Just remove the screw that holds the cover with the choke knob, and you have access to the foam filter, which you can clean and reuse.

FOAM AIR
FILTER ELEMENT

CHOKE PLATE

SLOT FOR
CHOKE KNOB STEM

CARBURETOR

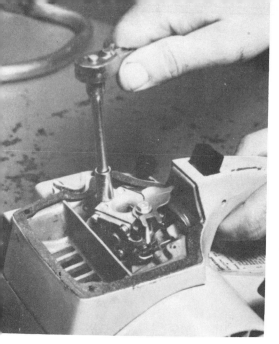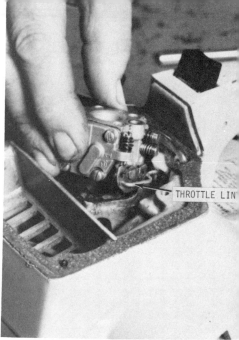

THROTTLE LIN'

5–130 and 131. Removing the carburetor is simply a matter of taking out two bolts, one of which is the bolt holding the choke plate, as shown left. Next, you just lift up the carburetor and disconnect the throttle link.

5–132. Lift the carburetor up still farther and you have easy access to the fuel line, which also must be disconnected to complete the job of carburetor removal. Replace the carburetor base gasket before refitting the carb.

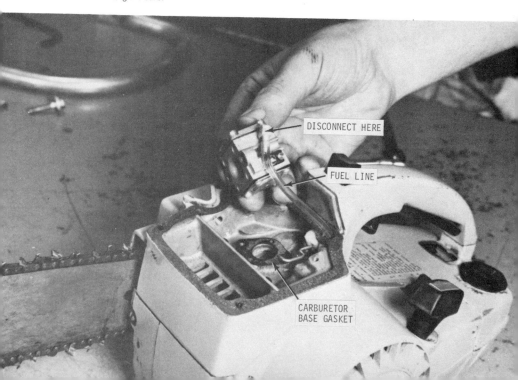

DISCONNECT HERE

FUEL LINE

CARBURETOR
BASE GASKET

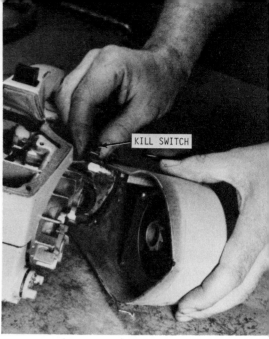

KILL SWITCH

5-133 and 134. Remove the flywheel-starter cover screws. Then, as the cover comes off, slip the magneto kill switch off.

5-135. Disconnect the wire from the magneto kill switch. For service to the ignition system, you need not strip the saw any further.

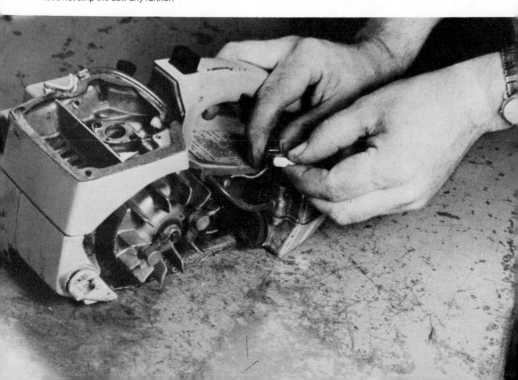

5–136. To go ahead with ignition service, remove the spark plug and replace or clean if necessary. Then stuff clothesline into the cylinder through the plug hole to serve as a shock absorber.

5–137 and 138. Hold the flywheel with one hand while you loosen the flywheel nut with a socket wrench. Do not remove the nut. With the nut projecting just above the crankshaft threads, as shown right, whack it with a hammer while you pry up with a screwdriver to break the flywheel loose from the tapered section of the crankshaft.

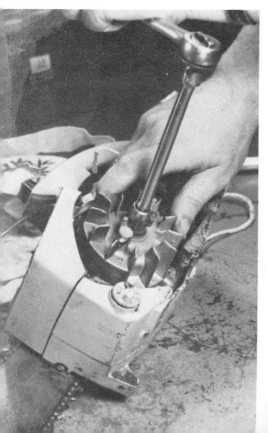

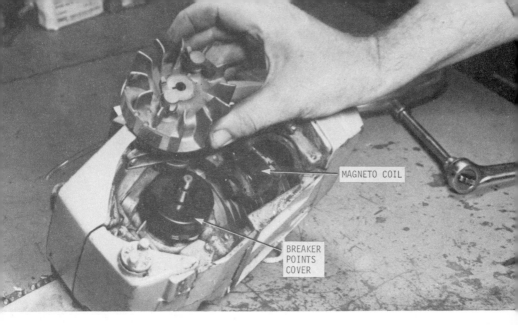

MAGNETO COIL

BREAKER
POINTS
COVER

5-139. Lift off the flywheel. Check the half-moon key and key slots in the crank and flywheel. Then remove the breaker cover to get to the points.

5-140. If the chain oiling system doesn't work properly, you must remove the breaker points plate and move the points and condenser aside.

BREAKER POINTS
PLATE SCREWS

BREAKER POINTS

CHAIN
OILER

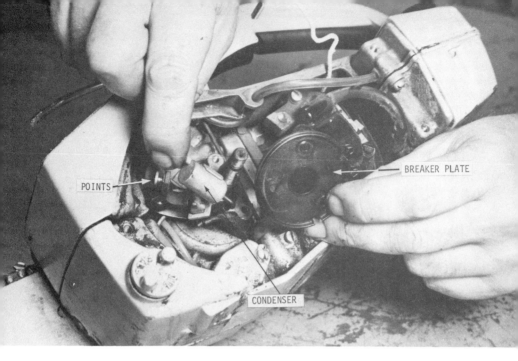

POINTS

BREAKER PLATE

CONDENSER

5—141. With the breaker plate out, move the points and condenser away from the oiler.

5—142 and 143. To provide additional working clearance, disconnect the wire from the points to the magneto coil, as shown left, and move it out of the way. Next, remove the two screws holding the oiler pump assembly to the engine.

OILER LINE
CLAMP

A

MOUNTING
SCREW
HOLES

5-144. The two mounting screws are out. Note that screw ''A'' is not a mounting screw and therefore does not have to be removed until you are ready to disassemble the pump. You must remove the oiler line clamp, however.

5-145. Remove the oiler line clamp with a screwdriver. Lift up the pump assembly and pull the line off the pump.

5-146. Here the pump is disassembled. Replacement gasket, diaphragm, springs and seals are normally included in a repair kit.

5-147. Remove the single bolt that holds the chain guide bar. This bolt also retains the chain cover. The two knobs on top of the cover are for the manual grinding stone mechanism that helps keep the chain from getting dull. Note the accessible muffler.

5-148. Inside the cover, you'll find the grinding stone mechanism. The part on the left connects to the adjusting knob, which regulates the position of the grinding stone to compensate for wear or to adjust for amount of grinding desired. Moving the handle knob as shown brings the stone in contact with the chain.

5-149. Use a drift and hammer to remove the clutch friction material locking hub, which holds the friction material assembly to the end of the crankshaft. Since the hub has a left-hand thread, whack it clockwise. The hub has an arrow and the word "off." Any shock to the engine's reciprocating parts, including the piston, is absorbed by the clothesline that fills the cylinder. Prior to removing the hub completely, check to be sure the clearance of the friction material from the circumference of the drum is uniform all around. If in doubt, check the clearance with a feeler gauge.

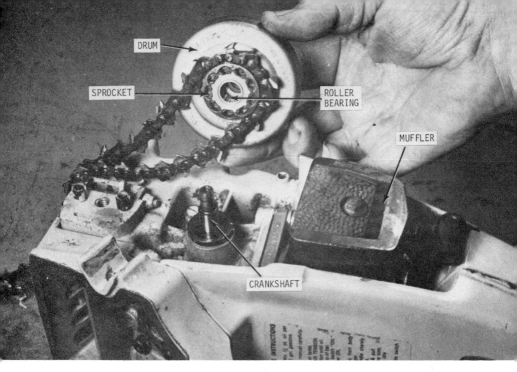

5-150. The hub is off. Lift the drum up and off and check the chain and sprocket for wear. In this case, the bearing has come off with the drum and sits inside. To check it, disconnect the chain; then refit the drum and spin it. It should spin freely and without wobble. When reassembling, coat the rollers with high-temperature grease such as Lubriplate.

5-151. Here's a close-up look at the pieces of friction material and the leaf springs that hold them away from the drum inner surface until centrifugal force from the spinning engine overcomes the force of the leaf springs.

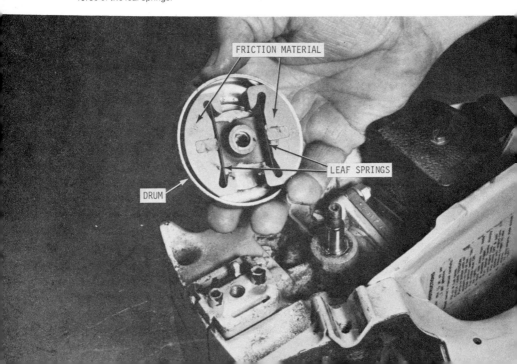

5–152 and 153. Muffler replacement is a simple remove-the-bolt job. Once the muffler is off, as shown at right, clean the exhaust port of any carbon accumulation, preferably with a wood scraper. If you use a screwdriver, be careful not to score the exhaust port surfaces or the piston.

5–154. Back to stripping the engine, remove the bolts holding the bottom cover.

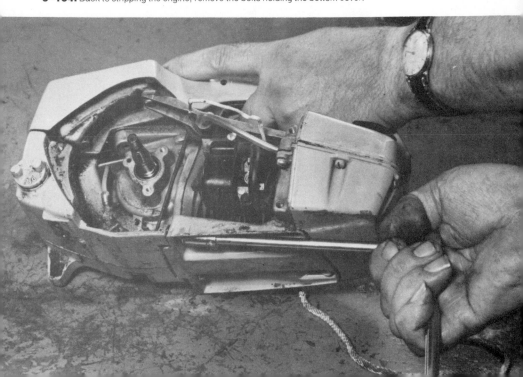

5—155 and 156. The bottom cover is off, at left. Its removal does nothing much for you except provide some additional access for a major cleaning of the engine. The top cover bolts are next. As you can see, at right, they're the Allen type (hex-head hole cut into a ground bolt head).

5—157. The top cover lifts off, completing the stripping of the engine.

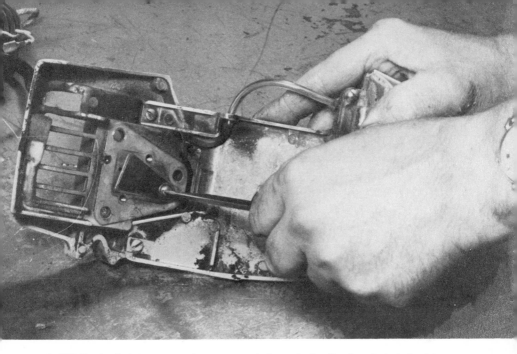

5-158. Turning the top cover over gives you access to the reed valve. Here the valve's single retaining screw is being removed.

5-159. The magneto coil screws are readily accessible when the engine is stripped, but so are they when just the flywheel cover is off. If you're bringing the engine in for a complete overhaul, however, you can save this piece of stripping for last.

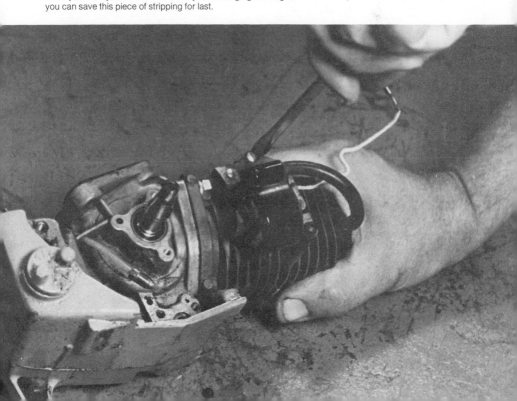

McCULLOCH TWO-CYCLE ENGINE

The McCulloch Mini-Mac 6 two-cycle engine can be removed completely from housing, as shown in the disassembly sequence (5–160 through 193). Some points to note include the following:

1. If turning down the adjusting screws as shown in 5–164 fails to provide clearance for removal of the housing, remove the screws completely.

2. If the engine doesn't come out as covered in 5–171, and you've moved the engine as far toward the guide bar side as possible, recheck the position of the flywheel fins. They must be positioned just right to enable them to clear the starter pawls.

3. Raised sections on the flywheel, as shown in 5–178 and 178a, permit holding the flywheel with a beefy screwdriver instead of the special spanner wrench.

5–160. This McCulloch Mini-Mac 6 has an engine that can be completely removed from the housing. Stripping it begins with removal of the air cleaner cover, held by the spring clip.

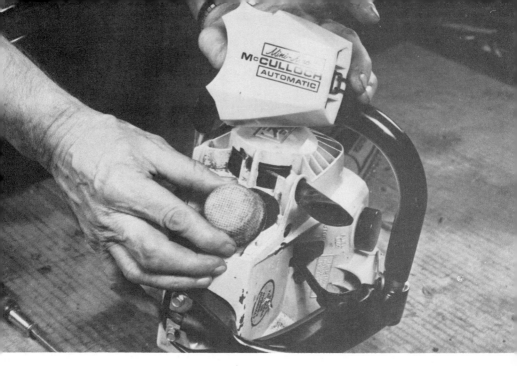

5-161. Lift the cover and you can pull out the air filter element for easy replacement.

5-162 and 163. Next remove the filter gasket, which should be replaced with the filter. Now turn the machine around, as shown at right, and begin the job of stripping it by pulling off the spark plug wire and, as shown, removing the choke knob, which is held by a single screw.

5–164 and 165. Turn down the idle-, low-, and high-speed adjusting screws so that when that section of housing is removed, it will clear the screws. To get a rough adjustment after reinstallation, you should count the number of turns on each screw, and make a sketch of the location of the screws relative to each other. As shown at right, the removal of the housing assembly bolts is the next step.

5–166. This bolt under the trigger is easy to forget.   **5–167.** Also often forgot is this underside bolt.

5–168. Lift out the oiler lever.

5–169. The front cover is the first part to come off, exposing the fuel and oiler tanks, which are in a single, divided canister.

FUEL, OILER
TANK
CANISTER

THROTTLE TRIGGER LINK

5-170. Disconnect the throttle trigger link, a tiny clip-like part. Just slip a screwdriver under it and pry gently.

5-171. Pull on the guide bar to move the engine to that side. It will move just a fraction of an inch, but this will be enough to get the flywheel starting pawls clear of the starter. Then turn the flywheel so that the fins line up in a way that permits the engine to come out.

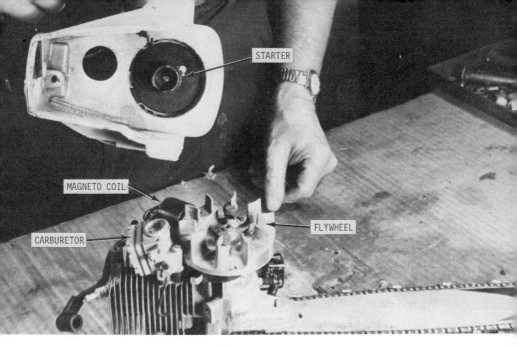

STARTER

MAGNETO COIL

CARBURETOR

FLYWHEEL

5-172. The cover is off. This gives you complete access to the flywheel. Once you remove the flywheel, you'll have easy access to the carburetor and ignition.

5-173. Removing the ignition system parts begins with pulling the coil's primary wire connector with needle-nose pliers.

5–174 and 175. Next, as shown left, take out the three screws that hold the magneto coil and carburetor. Then lift out the magneto coil.

5–176 and 177. Remove the carburetor, as shown left. Next, the "book" method of removing the flywheel is to use a socket wrench on the flywheel nut and a special spanner tool to hold the flywheel. Once the nut is off, just pry up the flywheel with a screwdriver.

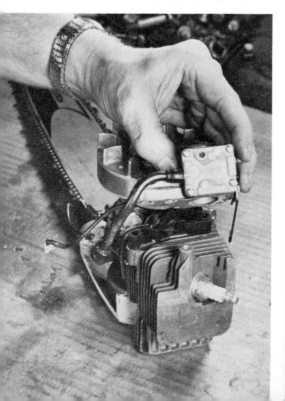

5–178 and 178a. A close look at the flywheel, however, shows that a special spanner wrench really isn't necessary, for there are raised sections of the flywheel against which a heavy screwdriver or something similar can safely bear, as shown above. The black parts on the flywheel surface, shown below, are pawls that engage the starter. The black cover on the end of the crankshaft covers the breaker points.

BREAKER
POINTS

5—179. Lift away the black cover to get at the breaker points.

5—180. Pull the half-moon crankshaft key with pliers.

5-181 and 182. Loosen the retaining screws. Then lift the entire breaker plate assembly out, as shown left. That's a muffler in the right-hand photo. It's held by two bolts.

5-183. Remove the muffler bolts. Once they're out, you can lift the muffler from the engine.

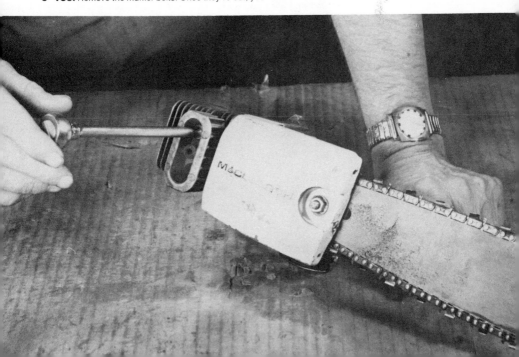

5—184 and 185. With a screwdriver, clean the exhaust port of carbon, as shown left. Be careful not to scratch the piston or gouge the port surfaces. Then with a socket wrench remove the chain guide bar bolt, which also holds the saw housing cover. Take off the cover, then the chain and guide bar.

5—186. Take off the hub that holds the clutch friction material assembly to the flywheel. This hub is on pretty tight, and you'll need a hammer-actuated manual impact wrench to loosen it. Stuff the cylinder with clothesline and leave the chain and bar in place to help hold back the clutch. Once it's loosened with the impact tool, it unthreads with a socket wrench. Before completely removing the friction material hub, spin the drum to make sure it spins freely and true. (If it doesn't, the bearing probably is defective.) Also check for even wear of friction material with a feeler gauge between the drum and the raised contact area of friction material. Clearance should be nearly uniform.

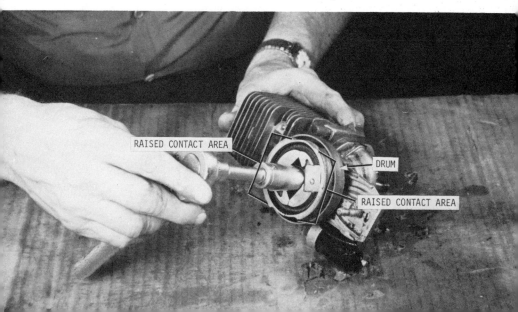

5-187. The friction material hub and drum are off, exposing the sprocket. In this case, the drum bearing remains on the crankshaft.

5-188. This engine has a combination manual and automatic chain oiling system. The screwdriver is on the automatic oiler's adjusting screw, which is threaded into the oiling pump assembly.

5-189. Disassembly of the oiling system begins with removal of the hose to the chain. It's being pulled with a needle-nose pliers.

5-190. Next, you should remove the manual oiler hose.

RETAINING CLIP

5-191. Flipping the retaining clip away permits access to the pump body.

5-192. Here the disassembled pump parts are (left to right) pump body, pump sleeve (a type of flow adjuster), piston assembly, piston return spring and the cavity in the engine. Note the hole in the pump cavity, which leads to the crankcase. In operation, the pump delivers in response to changes in crankcase pressure.

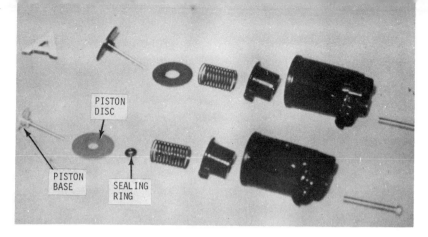

PISTON
DISC

PISTON
BASE

SEALING
RING

5-193. Can you tell the difference between the two disassembled pumps? Further, can you tell which is the pump shown in the previous illustration? The point of this photo is to demonstrate that you must eyeball parts very carefully in order to secure the proper replacements. The top pump is an older McCulloch design; the bottom one is the one shown in 5-192. The two additional parts you seem to see here (piston disc and ring seal) are already fitted to the piston assembly in 5-192.

5-194. WHY NOT TO TAKE IT COMPLETELY APART

Complete disassembly and overhaul of a small gas engine may not look difficult. But unless you have experience with one, or have done similar work on an automobile engine, you could easily get into trouble. This exploded drawing is of a McCulloch chain saw engine. Remember, a chain saw powerplant is far simpler than a four-cycle lawn mower or snow blower engine.

You may still think that nothing looks too difficult. Okay. The parts list below, keyed to 5-194, will acquaint you with some of the things you'll have to check. The list is not exhaustive, just representative.

Parts are as follows: 1) bolts holding the halves of the engine together; 2) lower engine half; 3) upper engine half, including cylinder which must be checked for scores and ridges; 4) crankshaft, the journals of which must be absolutely smooth, and neither out of round or tapered; 5) connecting rod; 6) piston with rings and connecting rod piston-pin bushing anmd bushing locks. In the circular inset to the right are 5A) connecting rod and one of two bolts that hold the lower end cap to the bearing; 5B) lower end bearing cap; 5C) two halves of connecting rod's roller bearing, which must be inspected and replaced if necessary; 6A) piston rings, which must fit to very precise tolerances both in the piston and in the cylinder; 6B) piston-pin bearing assembly, in which the connecting rod must be able to swivel freely, but without the least bit of looseness. Continuing now with the main drawing, the parts include the following: 7) crankshaft seals, which must be inspected and replaced if necessary; 8) bearing clip; 9 and 10) bearings, which must be inspected and replaced if worn; 11) crankcase cover; 12) cover gasket.

Among the tools necessary to completely overhaul even this simple engine (and beyond the scope of this chapter) are an arbor press, a pin driver, a piston support, a piston boss support, a bearing driver, a piston ring compressor. Trying to work without them, such as by driving the piston pin out or in with a hammer, would result in damaged parts.

Additional illustrations of some of the complications present in an overhaul are shown in 5-122 and 123, performed as an experiment during the disassembly of a Sears chain saw.

5-194

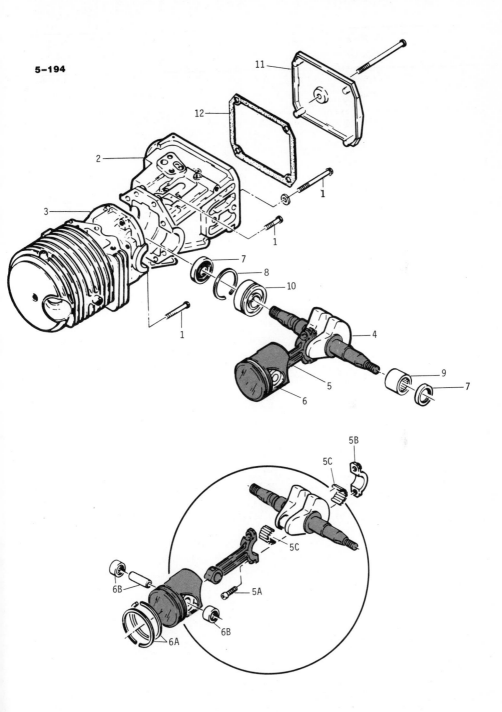

6

Servicing Ignition and Starting Systems

Even a careful reading of Chapter 1 might lead you to believe that the ignition and starting systems of typical small gas engine appliances require very little repair. After all, the only moving parts in the ignition system are the sturdy-looking flywheel with its permanent magnets and the breaker points; and the only replaceable parts are points, condenser and spark plug. The typical starter is a spring type with a pull cord.

Appearances can be deceiving, for the overwhelming amount of the service work on a small gas engine is devoted to the ignition system. See 6-1. Although the starter is not normally troublesome, if it does fail, there are some intricacies about its service you should know.

Ignition system service begins with the spark plug and its wire. There are ohmmeter tests for spark plug wire, but here are the practical approaches for the homeowner:

1. If the wire is cracked or oil-soaked, replace it.

2. If, when the engine is running or being cranked, you see sparks jumping from the wire to the engine metal, replace the wire.

3. If the plug wire terminal is loose on the plug, look inside the rubber boot on those plug wires so equipped. Perhaps the terminal can be tightened by squeezing gently with a pliers. If not, replace the wire.

4. If the magneto coil looks good, the flywheel magnets pass a test for magnetism, and the breaker points are properly adjusted, replace the wire if there is no spark to the plug (Magneto coil inspection and flywheel magnets' testing are described later in this chapter.)

5. If the plug wire is an integral part of the magneto coil, the number 3 check and part of check number 4 are academic.

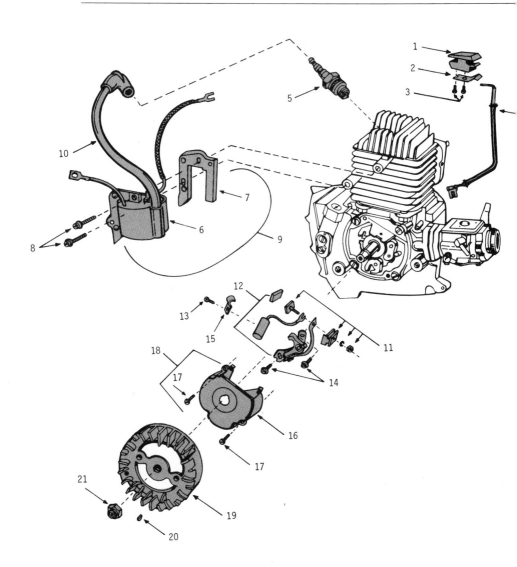

6—1. From this illustration you can see that although there are few moving parts in a magneto ignition system, there are many parts in all. The ignition system is the primary cause for servicing a small gas engine. The parts shown are 1) magneto kill switch; 2) switch leaf spring; 3) switch mounting screws; 4) switch wire; 5) spark plug; 6) magneto coil; 7) magneto core; 8) coil mounting screws; 9) coil and core as an assembly; 10) spark plug wire; 11) primary wire mounting hardware; 12) points, condenser and crankshaft lobe lubricating felt; 13) condenser mounting screw; 14) points mounting screw; 15) condenser clamp; 16) points cover; 17) cover screws; 18) points cover and screws as an assembly; 19) flywheel; 20) half-moon key; 21) flywheel nut.

THE SPARK PLUG

The spark plug itself is a much misunderstood item, and a lot of books have dwelt heavily on it. In fact, a reasonable discussion of the plug could fill a book all by itself. Rather than ply you with the kind of information of primary interest to an automotive scholar, we'll limit spark plug discussion to just what you should know to be moderately informed, so that you can buy the right plug and to install it properly. We covered the basics of spark plug operation in Chapter 1. Now let's take a closer look, referring to 6–2.

The high voltage electricity travels from the spark plug wire into the center electrode of the plug. When it reaches the bottom, it is supposed to jump to the side electrode, which is an electrical ground because it's part of the metal structure of the plug that is threaded into the engine's cylinder head. If the electricity jumps across the air gap of about .025 to .030 inch, it ignites the air fuel mixture and all is well.

Sometimes the spark won't jump the gap, in which case the engine won't run. The most common reason is that the ceramic insulator that separates the center electrode from the side electrode isn't functioning as an insulator, and instead is allowing the current to leak along the ceramic to an electrical ground. Without a current-jumping gap, you get no spark and no engine operation. Why would an insulator suddenly stop working? Here are the most common reasons: Coatings of oil, carbon (from a rich fuel mixture) and lead may form an easier path for the electricity than the gap between the electrodes. Or the gap between the electrodes may get sufficiently large so that even an incomplete coating on the ceramic insulator may be an easier path for the current. The spark plug fires thousands of times per minute and in time the electrode tips wear away, increasing the gap between them.

Electrode wear is a normal condition, and although the spark plug can be readjusted, it generally isn't worth the trouble. By the time the electrodes are badly worn, just a normal accumulation of lead deposits makes it worthwhile to replace the plug. Why not clean it? The only way to really clean a plug is with a sandblaster, which costs about $40, at least. But sandblasting is not recommended by some engine manufacturers, including Briggs and Stratton. New plugs are perhaps a dollar, so even if the manufacturer of your engine approves sandblasting, the equipment would be a poor investment that would take a couple of lifetimes to pay off in savings on spark plugs.

Thick carbon accumulations on the plug are abnormal. They indicate one of the following: 1) The clogged muffler (or exhaust port clogged on two-stroke engine) leaves exhaust residue in the cylinder. 2) The carburetor jet needle is adjusted for too rich a mixture. 3) The jet and/or needle is worn, allowing excess fuel into the mixture. 4) The choke is sticking closed, or the air filter is clogged, either of which can overenrich the mixture.

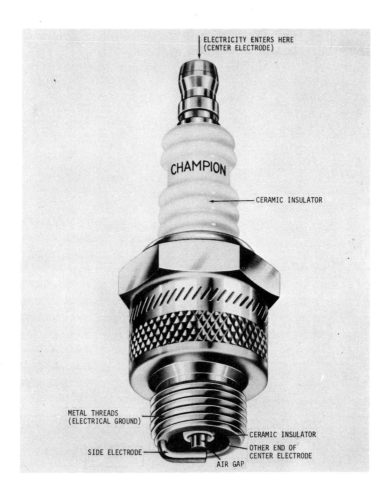

ELECTRICITY ENTERS HERE
(CENTER ELECTRODE)

CHAMPION

CERAMIC INSULATOR

METAL THREADS
(ELECTRICAL GROUND)

SIDE ELECTRODE

CERAMIC INSULATOR

OTHER END OF
CENTER ELECTRODE

AIR GAP

6−2. The spark plug may look more complex than it really is. It has no moving parts, and it merely serves as a terminal point for the high-voltage electricity produced by the ignition system. Following the path of least resistance, the current travels through the plug wire and enters the top of the plug by passing into a metal rod called an electrode, which is insulated by a piece of ceramic to hold the current in. The electricity travels down the rod to the end, then jumps the small air gap to the side electrode, which is a piece of metal bonded to the metal portion of the plug; when the plug is threaded into the engine, there is a continuous metal path along which the electricity can travel from the side electrode. The size of the air gap is important, for if it is too large, the electricity will not be able to jump across. Conductive deposits on the insulator also can prevent the spark across the air gap, by providing an alternative path from the center electrode that is easier to follow.

A certain amount of oil accumulation on the plug is normal on a two-stroke engine, abnormal on all but a very worn four-stroke. The reason is that in the two-stroker the oil is mixed with the gasoline to provide engine lubrication, so naturally some of the oil will get on the spark plug. If you mix in too much oil, you can expect oil soaking of the plug. But don't skimp on oil either, or the engine will burn out. The two-cycle spark plug is designed to operate with a fair amount of oil on the insulator. When it comes to a significant oil coating on the insulator, you've got to be practical, particularly with a four-stroker. The normal causes of oil-fouled plugs on a four-cycle are these: 1) too much oil in the crankcase, 2) wear of the piston rings, or 3) wear of the guides in which the intake and exhaust valves move. If the oil level is right and the engine runs for a reasonable amount of time before the plug becomes oil-fouled and requires replacement, you may be better off leaving the annoyance rather than spending money for an engine overhaul. It takes just a few minutes to replace a plug, the price is low, and there are worse things to worry about.

REPLACING THE PLUG

The rules for plug replacement are simple:

1. The new spark plug should look almost exactly like the old from the hexagonal section down to the electrodes, assuming you got the right plug in the first place. The length of the threads should be the same and the amount of air space between the insulator and the metal structure should be similar. Why this warning when you can assume the parts supplier will look up the number in a book? There are two answers: You may not have the information on your engine necessary to identify the catalog listing. And even if you bring the old plug with you and get another brand's listed equivalent, the catalog listing may be in error.

2. Stick to name brand plugs, which are more likely to have a spark plug number and design that are closest to the original item. The off-brands may be different in design characteristics you can't see. Although they may work, they may not work as well.

3. The plug wrench you use is almost immaterial, because the small gas engine plug is wide open, or else you must remove a sheet metal cover to get at it. And once the cover is off, the plug is wide open. If you have a ratchet and deep socket, fine. Otherwise an inexpensive plug wrench will do.

4. Apply steady force to loosen the old plug. If it doesn't break free, squirt penetrating oil down onto the edge of the hole; allow it a few minutes to work its way in; then try again with the wrench.

5. If you're installing a new plug, round-wire spark plug feeler gauges (6–3) are best, but breaker point feelers will do. To increase the gap, pry up the side electrode, using a small thin screwdriver braced on the base of

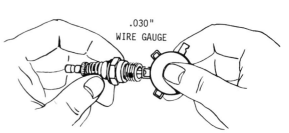

.030"
WIRE GAUGE

6–3. You should measure the gap between the spark plug electrodes with a gauge, preferably a round wire type. The specified gap for Briggs and Stratton engines is .030 inch. For most other small gas engines it's .025 inch.

the plug. To decrease the gap, gently tap the side electrode with a small hammer or a rock. The specified gap for most small gas engines is .025 inch, with Briggs and Stratton an exception at .030 inch. The store selling the plug should have a chart you can check. If the set of feeler gauges you have does not include the appropriate thickness, use two or more. Example: A gauge of .016 inch and another of .014 inch together equal the .030-inch thickness for a Briggs engine.

6. If the spark plug has a gasket, thread it in until finger tight, then with the wrench one-half turn more. If it hasn't got a gasket, thread in till finger tight, then just enough more until it feels tight. This may be as little as ⅟₃₂nd *of a turn, so don't let the absence of the gasket fool you into overtightening. The plug has a tapered seat and seals very nicely with relatively little force on the wrench. See 6–4.*

It's possible to do other ignition work on the mower and blower with the engine on the chassis; but this is not always a good idea. You would have to brace the engine and chassis on its side in the case of the typical reel mower and blower, and a crack in the chassis might result if the chassis slipped off and hit the concrete floor of your garage. With a rotary mower, the ignition system is at the top and in-chassis service is perfectly acceptable. On chain saws, in-chassis service is the only way on most, and it's practical on all.

Removing the engine where possible and servicing it on the work bench, even if not absolutely necessary, does have its advantages. You generally have your best lighting at the work bench, plus a vise, and you can work in the convenient standing position.

Begin by disconnecting the spark plug wire from the plug. Cover the terminal of the wire with electrical tape to prevent accidental firing. Take the metal cover off the flywheel. It's easy enough to identify because the starter spins the flywheel, the cover containing the starter is the one that covers the flywheel. You don't have to remove the starter from the cover.

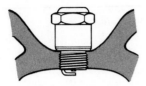 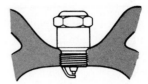

6-4. The amount of tightening-down required for the tight seal of a spark plug depends on whether or not it has a gasket. The left-hand drawing shows a plug, with gasket seated in a flat surface. The right-hand drawing shows how a plug without a gasket forms a tight seal by means of a tapered seat. The plug with a gasket may require half a turn (180 degrees) from the hand-tight position to seat. The tapered seat plug will be adequately tight with just a little turn past the hand-tight position.

If you have a mower or a blower with an electric starter, the starter must engage the flywheel. So locate the starter, and there you'll find the flywheel and cover.

The metal cover normally is held by screws of the slot or Phillips-head variety. If the screws are also hex-head and are unusually tight, use a socket wrench to remove them.

The exact side-cover removal procedure varies according to appliance and engine. Refer to Chapter 5 for guidelines and some examples. In a few cases, it may be necessary to remove some other parts and covers first.

Once the cover is off, you'll see the screen guard for the flywheel. (Or it may have been at least partly visible before you removed the cover, another tip-off to the location of the flywheel.) This screen prevents pebbles from damaging the flywheel's fins, which serve as a fan, aircooling the engine. The chain saw does not have the screen guard.

Turn the flywheel slowly by hand to make sure that it at no point comes in contact with the magneto coil assembly. If the two parts touch at one or more points, there is a strong chance that the magneto coil has been damaged. And if you are tracing a no-spark problem, you probably have found it.

If the magneto coil is tight on its mounting screws and the gap seems to be adequate, leave it alone. Many coil mounting screws go into elongated holes, so that if the gap is inadequate, the screws can be loosened and the coil moved away from the flywheel to avoid contact. Do not move the coil any farther away than specified, or the magnets may not permit sufficient magnetic induction of current through the coil. In any case, do not disturb the adjustment until you're familiar with the possible effect on ignition timing, covered later in this chapter. Happily, most magneto coils are simply screwed down into position, and you can't disturb them.

If you wish to check the gap between the magnets embedded in the outer circumference of the flywheel and the coil surface just opposite, use a brass or plastic feeler gauge. You really shouldn't use an ordinary automotive feeler gauge because it is made of steel, which will be drawn to the flywheel magnets and will give a false sensation of drag. Instead use a brass gauge (available from automotive supply stores) or a plastic gauge (sold by some small gas engines parts outlets), either of which is nonmagnetic. See 6–5.

Note: On those few flywheels with ring gears, the coil magnets may be inside the flywheel, because the gear teeth on the flywheel exterior may prevent location of the magnets there. The only way to check this internal gap is to slip the feeler in place as you temporarily reinstall the flywheel. Then turn the flywheel to feel if it turns with moderate drag.

Also check the flywheel or motor magnets for strength. They cannot be

6–5. The gap between flywheel magnet and magneto coil lamination should be measured with a feeler gauge, preferably one of plastic or of nonferrous metal such as brass.

remagnetized. If they have lost their magnetism, the flywheel must be replaced. A simple check is to hold a lightweight screwdriver by the handle, with the tip ¾ to an inch away from each magnet. The magnet should attract the screwdriver from this distance. See 6-6.

Now proceed to remove the flywheel, which is admittedly not the easiest job on most small gas engines. We covered most of the basics of this job in Chapter 5, as part of the disassembly sequences. We'll now review the subject in some additional detail and with illustrations of flywheel removal procedures that differ from those for the engines disassembled in Chapter 5. Where appropriate, references to illustrations in Chapter 5 are given.

Except on Briggs and Stratton, and most chain saws, flywheels are held to the crankshaft by a combination of retainers: When a nut is tightened, it pushes the flywheel onto a tapered section of the crankshaft. This tapered section retains the flywheel even after the nut is removed. Then a half-moon shaped key fits into the recesses in both the flywheel hub and the crankshaft to keep the flywheel from turning. If there is any wear in the key or in the recesses into which the flywheel fits, the flywheel vibrates; this can cause several problems, namely these: 1) The position of each

FEELER GAUGES

If you have never used a feeler gauge before and don't know what it is, be assured it's nothing sophisticated. A feeler gauge is a strip of material of precise thickness for measuring the gap between two parts. A gauge of appropriate thickness is pushed in between the parts, and if it passes through with just moderate drag, the gap between parts is equal to the gauge thickness. If the gauge won't pass, the gap is smaller. Then you must check with a gauge that is a few sizes thinner. If it slips in easily, try one a little thicker. All feeler gauges have the thickness marked on one side.

In most cases the gap must be precise, and you may have to make adjustments. The gap between the flywheel and magneto coil is not so critical, however, so long as it is equal to or in excess of the specified minimum. In the absence of the specifications, hope for a gap of at least .008 inch and preferably .010 inch. The Briggs Magna-Matic is a notable exception; it has a .004-inch minimum.

If you have steel automotive feeler gauges and don't want to go through the expense of buying other types, here is a way of saving money: Slacken the magneto coil mounting screw(s) and move the coil away from the flywheel. Turn the flywheel until a magnet in the circumference is directly opposite. Insert the appropriate feeler between them; then the magnet should draw the feeler and the coil to it, setting the coil position correctly. Just tighten the coil mounting screws and remove the feeler.

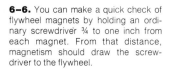

6–6. You can make a quick check of flywheel magnets by holding an ordinary screwdriver ¾ to one inch from each magnet. From that distance, magnetism should draw the screwdriver to the flywheel.

1"

magnet relative to the magneto coil varies slightly, changing the time that current is induced in the coil and the time the spark arrives at the plug, which can noticeably affect engine performance. 2) The flywheel vibration accelerates wear on the key and its recesses, and soon makes it necessary to replace the flywheel and crankshaft. Checking the condition of the key and recesses, therefore, is a very important aspect, and we'll discuss how to do it a bit later in this chapter.

To remove the flywheel, remove the nut that holds it. This means you must hold the flywheel to keep the crankshaft from turning while you put wrench pressure on the nut. You could try to wedge a screwdriver between the fins of the flywheel to restrain it, a procedure that will work nicely on most McCulloch chain saws because there is a raised shoulder for the screwdriver at a couple of opposite points next to the fins; see 5–178 in Chapter 5. If you're lucky, the flywheel nut is lightly tightened, such as on the Sears Explorer II, and you can hold the flywheel with your hand, or shock the nut loose as shown in 6–7 through 6–9. However, on other engines, you can't hold it by hand; and if you use a screwdriver wedged in, the result could be broken fins, and you still might not be able get the nut off. A replacement flywheel could cost you up to $20 or more. (Note: If you insist on using a screwdriver and break a fin, then intentionally break off the fin directly opposite—180 degrees away—to minimize the amount of flywheel imbalance.) These broken fins will result in a small loss of cooling air circulation, and the remaining imbalance still might be sufficient to measurably shorten the life of the engine. But if you tried the screwdriver procedure, you're obviously a gambler, so why stop suddenly.

6-7 and 6-8. A typical flywheel removal job begins with removal of the flywheel nut, as shown left, shocking the nut loose with hammer blows on a ratchet. Note that the flywheel has a ring gear in its circumference. Once the nut is off, thread on this closed-end nut as shown at right (or two ordinary nuts, with the second locked to the first and projecting just above the tip of the crankshaft). Whack the closed-end nut with hammer. And with a screwdriver, pry the flywheel loose from the tapered section of the crankshaft.

179

6-9. The flywheel is off. Although the flywheel removal procedure is similar to that described earlier, the layout of the ignition system itself is a bit different. Flywheel magnets are on the inside surface, as is the magneto coil. The breaker points are in their usual location.

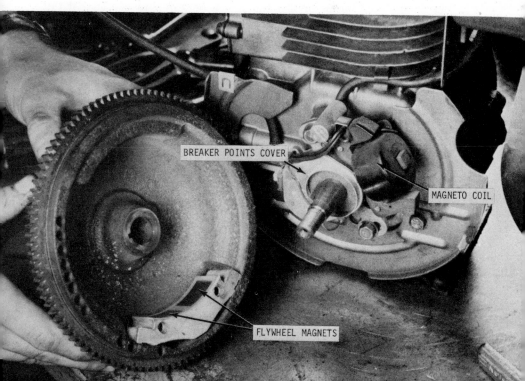

BREAKER POINTS COVER

MAGNETO COIL

FLYWHEEL MAGNETS

The better, safer ways are these:

1. Use a manual impact tool, sold in most auto parts stores. See 5–79 in Chapter 5. Unlike the professional impact tool powered by electricity or by compressed air, this one is whacked with a hammer and it shocks the nut, bolt or screw loose better than the ratchet and hammer setup in 6–7. The tool has many household and automotive uses. Suggestion: To prevent possible damage to the piston, remove the spark plug, turn the flywheel until the piston is down, and thread clothesline through the spark plug hole into the cylinder to serve as a shock absorber.

2. Get a flywheel holder from the parts distributor. As shown in 5–62 in Chapter 5, it allows you to restrain the flywheel by grabbing both ends of two fins. Although limited in scope, it's cheaper than the manual impact tool ($3 to $6 versus about $12).

3. Try to jerry-rig something that will hold the flywheel without damaging it. If the flywheel is small enough and the nut isn't impossibly tight, an automotive oil filter wrench might fit around and hold it. A chain vise-type plier wrench also might work. Either might require you to remove the magneto coil to get the wrench all the way around. So if you do, be sure to scribe alignment marks on the magneto and engine with touchup paint, nail polish or the like. See 6–10.

On most Briggs and Stratton engines, there is no flywheel nut. A special clutch assembly for the starter has a threaded hub that performs the same function. There are special tools that hold the flywheel and enable you to loosen the hub, but a soft flat-end punch or drift whacked against the

---

6–10. If ignition timing has been disturbed, or if you must remove the entire ignition assembly for service of other parts, make sure there are alignment marks to insure proper ignition timing. In this photo, there is a mark on the bolt hole boss, but none on the breaker plate, so a mark is being scribed with a screwdriver. The breaker plate hole is elongated for adjustment. So without the mark, there would be too great a timing variation possible.

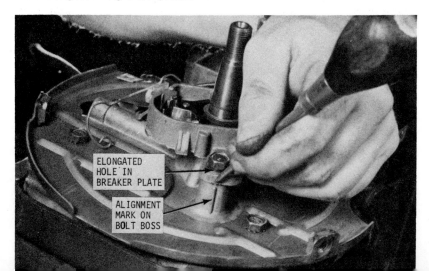

clutch ears with a hammer usually will do the job as well; see 5–62 and 63 in Chapter 5. If the Briggs engine has a nut, see 6–11.

On all but the Briggs setup and many chain saws, the next step is to shock the flywheel loose from its tight fit on the tapered part of the flywheel. Most professional shops use a special closed-end nut and a hammer (6–8). They thread the nut onto the flywheel, until it is finger tight, and then whack the closed-end top with a hammer, while prying from underneath the flywheel with a screwdriver or two.

The closed-end long nut is called a shock nut, and although it's not expensive and is reasonably available, you can achieve the same effect by just loosening the flywheel nut and threading a second nut onto the flywheel, so that the second nut projects above the end of the flywheel. With wrenches, lock the two nuts together, and you've got the equivalent of a shock nut.

Many chain saws do not have the heavily tapered flywheel end, so that once the nut is off, gently prying with a pair of screwdrivers will lift the flywheel up and off.

If you're lucky, somewhat more than gentle prying with screwdrivers will lift the Briggs flywheel; see 5–66 in Chapter 5. If this doesn't work, you probably won't be able to use the shock nut system because the end of the typical Briggs flywheel isn't threaded. Briggs does make special pullers (6–13 through 6–14) to cover the situation. Tecumseh also has one (6–15), but the Tecumseh's shock nut works every bit as well.

Although Briggs and most chain saw engines have flywheel retention devices that differ from the usual in small gas engines, all engines have flywheel keys. So the key deserves your special attention.

CHECKING THE FLYWHEEL KEY

Once the flywheel is free of its fit on the tapered portion of the crankshaft, try to move it clockwise and counterclockwise without turning the crank. If it moves even a few thousandths of an inch, check the half-moon key that positions it on the flywheel and the recesses into which the key fits (6–16). Normally the wear occurs in the key, and a new key is an inexpensive piece of insurance if you have any doubt. If the new key fits sloppily in the flywheel, the flywheel must be replaced. If it fits loosely in the crankshaft, the engine may require overhaul, and you might consider a replacement engine, as discussed in Chapter 9.

THE IGNITION POINTS

Once the flywheel is off (6–17), you'll see the ignition points or the sheet metal or plastic cover under which the points are located. The points and

BRIGGS AND STRATTON FLYWHEEL REMOVAL

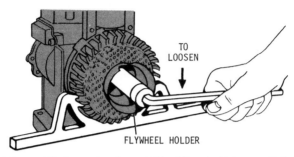

6–11. Many Briggs and Stratton engines have a left hand thread, which means that clockwise pressure must be used to loosen, as shown in this illustration. In general, the flywheel with a half-inch nut has a left-hand thread. The flywheel with a ⅝-inch nut has the conventional right-hand thread.

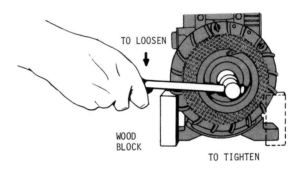

6–12. On Briggs flywheels of greater than 6¾-inch diameter, a flywheel spanner wrench isn't even available, but fins are beefy enough for you to hold the flywheel with block of wood, as shown.

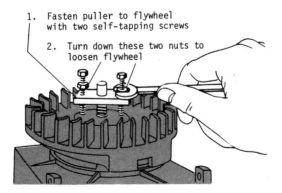

6–13. If a Briggs flywheel has two holes provided. you can use a puller, as shown.

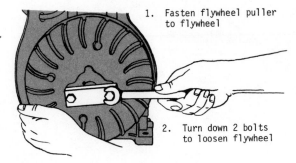

1. Fasten flywheel puller to flywheel

2. Turn down 2 bolts to loosen flywheel

6–14. This is another type of flywheel puller for Briggs engines, but it's designed for those with Magna-Matic ignition. With this ignition, you probably won't have to pull the flywheel as often, for points and condenser are outside the flywheel. The magneto coil, however, is under the flywheel, so if you want to adjust the coil-to-magnet gap, you'll have to take the flywheel off.

condenser setup looks very much like that in a pre-1975 automobile arrangement; but there are a number of differences, both in operation and service. See 6–18.

In the automobile, the points and condenser are housed in a distributor, which also has the job of transmitting spark to the correct plug. With a one-cylinder small gas engine, there's only one correct plug, so there are fewer parts.

Whether the engine is a two-cycle or a four-cycle, the points are opened and closed by the crankshaft in all but a few four-cycle engines, such as the Briggs MagnaMatic, in which the points are operated by the camshaft (as in the automobile).

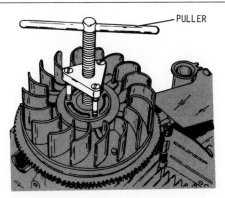

PULLER

TECUMSEH PULLER

6–15. This flywheel puller is for a Tecumseh engine, but on this one the use of a closed-end nut (or two regular nuts) also works. Thus, the tool is strictly a luxury item that makes the job a bit easier.

FLYWHEEL RETAINING
KEY IN CRANKSHAFT

6-16. Checking the half-moon key that retains the flywheel to the crankshaft is very important part of ignition service. If the key is worn, replace it. If the key has enlarged its slots in the crankshaft or flywheel, these parts should be replaced. The flywheel is easy to replace, but the crankshaft isn't, so if the enlargement is in the crankshaft slot, you may be better off replacing the engine. An alternate possibility is to take the engine to an automotive machine shop, where an automotive half-moon key might be cut down to fit in tightly.

6-17. Lift off the points cover (in this case held by a spring clip). Now the entire ignition system is accessible for service.

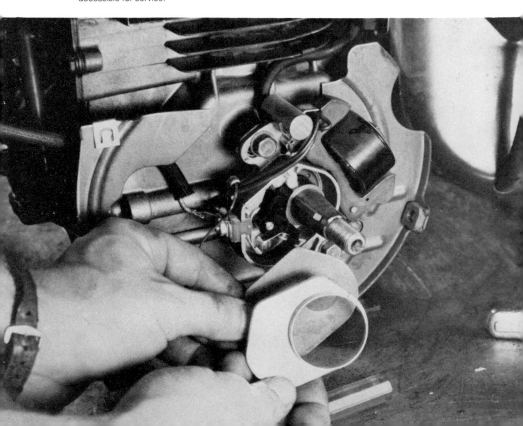

A small lobe on the crankshaft (or camshaft) pushes one of the breaker points away from the other, which is fixed in position. The push may be direct, or indirect. In the indirect setup, the crank lobe pushes on a short rod called a plunger, which pushes on the movable point. As the crank continues to turn, the lobe releases pressure and a leaf spring brings the movable point back into contact with the fixed point. See 6–18 and 21.

The gap between the two points when the lobe is at the maximum push position is extremely important. If the gap is too small, current can arc across it, and the break in the circuit that is necessary to create the spark will not occur. If the gap is too great, the points will open a bit earlier than they should and close a bit later. This can have two consequences: Inasmuch as the opening of the points results in the spark, the plug will fire too early, reducing performance. Or the late closing of the points may reduce the electrical charge absorbed by the magneto coil, weakening the resulting spark.

KILL SWITCH

To stop the small gas engine with the magneto ignition, a kill switch is used (Chapter 5: 5–57, 76, 99). It's normally a button near the throttle on chain saws or perhaps a small lever on top of the mower or blower engine, in either case labeled "STOP." A wire leads from one terminal of this switch to the magneto coil. The other terminal is mounted on the engine, which is an electrical ground. When you push the button or pull the lever, the two terminals are brought into contact, creating a short circuit that prevents the magneto from functioning, thus stopping the engine. When an engine fails to stop after the kill switch is activated, the possible causes are these:

1. The wire to the magneto has come off either the switch contact or the magneto. In the first case, the wire end may or may not be touching the engine. If it touches, it shorts out the magneto and the engine won't start. If the engine starts and vibration moves the wire, allowing the end to touch the engine, the engine will stop.

2. The switch contacts are not coming together when the kill switch is activated. In many cases, this can be cured by removing the cover plate that holds the switch, observing its action as you operate it, then bending a contact tab on the terminal with the wire from the coil so that the tab makes contact when you close the switch. If there is no external tab, the switch must be replaced.

When an engine fails to start, the kill switch also may be responsible. In this case, the switch may be short-circuited, preventing the magneto from functioning just as if you turned the switch to "off." To check this as a cause of starting failure, disconnect the wire from the switch and check for spark to the plug, as explained in Chapter 4. If disconnecting the kill switch wire cures the no-spark problem, replace the switch.

SERVICE

With the flywheel and breaker points cover off, you can see the points and condenser. (On many saws the condenser is not with the points, but outside

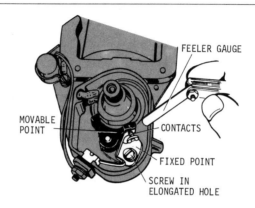

6—18. This shows the typical breaker points setup with the elongated hole in the fixed point plate. When the points are open to maximum possible, insert a feeler gauge of specified thickness between the two contacts.

6—19. This close-up look at the contacts of the breaker points shows two types of metal transfer. On the left, metal has been transferred from the fixed point to the movable one, indicating that the condenser is of low capacity. In the right, the metal has been transferred from the movable point to the fixed, indicating that the capacity of the condenser is too high. In either case the condenser should be replaced. If the points have an even gray coat on each contact face, the condenser is in good condition and need not be replaced with the points.

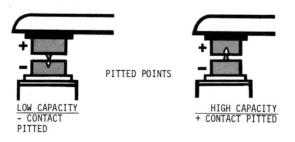

the flywheel near the magneto coil.) Before you take any action with screwdriver, wrenches and pliers, make a careful visual inspection. Is there an oil film all over? See the discussion of this subject later in this chapter. How is everything assembled? Observe carefully so you can reassemble. See 6–18 and 21.

Now turn the crankshaft clockwise until the breaker points open all the way. In most cases the movement of the points as you turn the crankshaft by hand is so gradual that it's difficult to actually see when they are opened to the maximum. A set of automotive feeler gauges can help. As soon as you see the points open, stop turning the crank and insert different size feeler gauges between the contact faces until you find one size that fits in with light to moderate drag. This is a starting point.

Next turn the crank just a tiny fraction of an inch more to see if a thicker gauge will now fit between the points, and repeat. When a couple of tiny movements of the crank no longer increase the gap, you've found the maximum. If you overshoot, turn the crankshaft back.

If the points remain closed no matter how much you turn the crank, one of the following things may have happened:

1. The locking screw or screws for the fixed point has slipped. In the case of the Briggs engine, the fixed point is built onto the condenser, and the condenser is held by a clamp that has loosened and permitted slippage.

2. The movable point's fiber block, which is the part that bears against the crankshaft, has worn.

3. In the case of the Briggs engines with the plunger between the crankshaft and the points, the plunger is worn.

By comparing old points with a new set, you can determine if the rubbing block has worn down. The plunger design, however, requires removal of the points. In either case, the failure of the breaker points to open will prevent the engine from starting, by eliminating the spark to the plug.

Before removing anything, look at the points' contact faces. Brand new, they are clean and shiny. Some new sets come with a coating of preservative, which can be removed with gasoline and a lint-free cloth to expose the shiny clean surface. In normal use they become light gray. If you see burn marks, or a transfer of metal (tiny metal hill on one, little valley in the other), the points and condenser should be replaced. See 6–19. Never file point faces clean except in a pinch. Filing will clean the surfaces and make them electrically conductive again, but small particles from the file will become embedded in the faces and the faces will quickly burn again. A filing might clean them sufficiently for one or two jobs, however; so if you can't get replacement points and must do something quick, go ahead. An ordinary nail file can be used, but it's rather coarse. There are special files made for emergency dressing of breaker points, and they are inexpensive.

REPLACING THE BREAKER POINTS

These are the important things to understand about adjustment and replacement of breaker points:

1. All point gap adjustments are based on moving the fixed point away from the movable one, by slackening the screw that normally keeps it stationary, and doing one other thing to reposition it. That one other thing may be something as simple as pushing the fixed point away with a screwdriver, turning a screw-type adjuster or, in the case of the Briggs, moving the condenser with the fixed point on it away from the movable point. Always lock the fixed point screw firmly when the gap is correct. (If moving the fixed point and not touching the movable one sounds contradictory, it isn't. The movable point has a designed-in range of travel, determined by the size of the crankshaft lobe and/or plunger. The only way to change the maximum gap between it and the fixed point, therefore, is to move the fixed point away or toward it.)

2. The point gap is always measured when the points are at the maximum opening, that is, when the tip of the lobe on the crank or camshaft is bearing against the movable point's fiber block or the plunger.

3. The method of adjustment of the fixed point varies according to engine manufacturer and even among different models by the same manufacturer. The simplest arrangement is a lockscrew in an elongated hole, or the Briggs setup that has a condenser with fixed point in a clamp (6–27).

4. If the breaker points are completely closed, no matter how you turn the crank, you'll have to determine the reason. As explained earlier in this chapter, the most common reasons include wear on the fiber block or plunger, or a loose fixed-point lockscrew (or Briggs condenser clamp). But there are others, including 1) wear of the crankshaft; 2) worn threads on the lockscrew or breaker point plates, permitting the fixed point plate to move from the shock effects of engine vibration.

To check these possibilities, begin by slackening the lockscrew and moving the fixed point plate an arbitrary distance away from the movable point. Retighten the lockscrew. No precision is required; just make a gap you can fit a feeler gauge into. Then turn the crank and keep checking to see if the gap gets larger or smaller, and if it does, there's hope. When the gap is at its largest, set it for .020 inch and tighten the lockscrew. Turn the crank and see if the points close all the way (so they touch) and then reopen to .020 inch. If they do, the crankshaft lobe and on the Briggs engine, also the plunger, can be presumed to be in acceptable shape. If you have a micrometer, you can remove and measure the Briggs plunger, but that's book stuff. The plunger is easy enough to just pull out when the points are removed, but all you really have to do is look at it and make sure the tip that bears against the crank lobe doesn't look worn. If it looks worn, replace it. Be sure to reinsert it correctly. If it goes in the wrong way, engine oil will leak onto the points, ruining them. See 6–30.

If crank or cam lobe seems to be doing the job and the point rubbing block isn't worn, you still can't assume the breaker point plate lockscrew threads are bad. First of all, the problem could be the screw itself, so why not just replace that and try the engine out for a while. Second, threads do not hold forever, and if it's been some time since the points last were adjusted or replaced, engine vibration and resultant breaker point movement could have brought the points together. You only go for a breaker plate replacement when the problem recurs just a few hours after the adjustment. Even then, you should first try a new lockscrew (and lockwasher if used) and make sure you snug the screw down tight. You can also try coating the screw threads with Loctite, a fluid that helps retain tightness.

THE ACTUAL WORK

When the crank is turned so the points are open to the maximum gap, the remaining work is relatively simple. See 6–20 through 34. Just loosen and remove the lockscrew (or two lockscrews on a rare engine) that holds the points. On Briggs engines with the point on the condenser, slacken the lockscrew and slide the condenser out.

You also will have two wires to disconnect—one from magneto coil to movable point terminal, another from the condenser to the movable point terminal. (The Briggs engine is an exception, with only one wire to the fixed point on the condenser.) Normally these will be screwed-down terminals, but on the Briggs engine a spring-loaded type is used at the condenser. To release it, you push in the end cap and pull the wire down and out. Briggs' breaker points sets come with a plastic piece that fits over the end cap, so that your finger pushes on soft plastic instead of sharp metal.

Install the new points, making very sure that the movable point has a fiber washer between it and the bottom of the pivot over which it fits. This washer may be factory-installed, or you may have to install it yourself. If you don't install it, the movable point will be grounded at all times, and the ignition system won't work. (The Briggs engine again differs, for the movable point is grounded, not the fixed one; hence no fiber washer.) A good rule to go by is this: The manufacturer is too cheap to put anything in his parts box that isn't absolutely necessary. So if it's there, it has an important function.

The points should make good face-to-face contact when closed. If necessary, bend the tab holding the fixed point to insure alignment. You can buy an inexpensive tool for this purpose from any auto parts store, but careful

BREAKER POINT REPLACEMENT

6—20. Let's begin breaker point replacement on a Tecumseh chain saw engine (two-stroke-cycle) with removal of the two cover-retaining screws. (The flywheel has already been removed.)

6—21. With the cover off, you can see points, condenser and terminal for wiring that leads to the magneto coil. Note that there are two screws, both in elongated holes, for the fixed point.

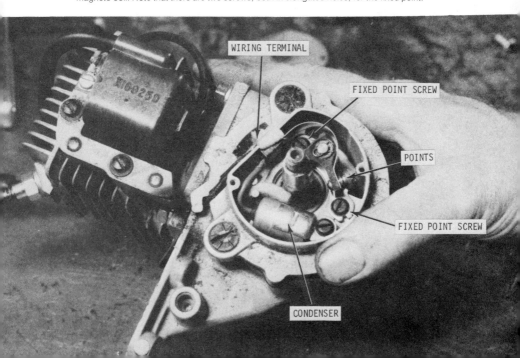

WIRING TERMINAL

FIXED POINT SCREW

POINTS

FIXED POINT SCREW

CONDENSER

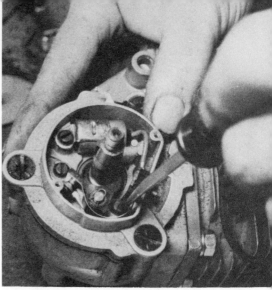

6–22 and 23. Points and condenser replacement begins, left, with unthreading the nut that holds the wiring to the terminal. Then, as shown at right, remove the two screws that hold the fixed point plate.

6–24 and 25. As shown below left, lift up the terminal from which you took the nut, and the points will come up and out too. If you want to leave the condenser in place, disconnect the condenser wire from the terminal. If you wish to replace the condenser too, remove its retaining screw. When installing new points, below right, turn the crankshaft until the points are wide open. Then insert a feeler gauge and adjust the gap for moderate drag on the feeler by moving the fixed point plate with a screwdriver. When the gap is correct, tighten the fixed point plate screws.

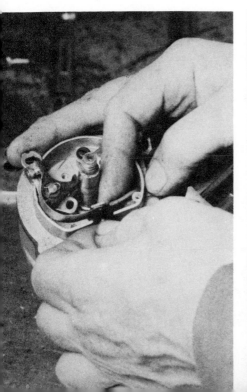

6—26. Replacement of the points and condenser on the Briggs engine is a bit different from that for other makes; although it starts out the same, with removal of screws that hold the points cover.

6—27. Briggs has a fixed point built into the tip of the condenser and a coil spring for the movable point instead of the conventional leaf spring. And instead of an exposed high point on the crankshaft, the crankshaft is covered, and the high spot on it pushes against a plunger rod. Also observe that the movable point and its pivot are all metal and that the wire is connected to the part of the condenser with the fixed point. In the Briggs system, the fixed point is not grounded, and the movable point is. The effect is the same, however, for when the points are closed, there is a complete ignition circuit from the magneto coil through the fixed point into the movable point, which is an all-metal part in contact with the engine metal, providing the electrical ground that completes the circuit.

6–28. Point replacement begins with turning the crankshaft until the point gap is at its widest, as measured with a feeler gauge. Then, as shown, remove the nut that holds the movable point to its pivot post.

6–29 and 30. Below left are the parts of the movable point. They are, left to right, the coil spring, the point and the pivot. Notice the wire from point to pivot, which provides a positive metal-to-metal connection to insure proper grounding of the part. Next, as shown below right, withdraw plunger with fingers and check for wear. If you have a micrometer, the length of the plunger should be at least 0.871 inch. If it is less, replace it.

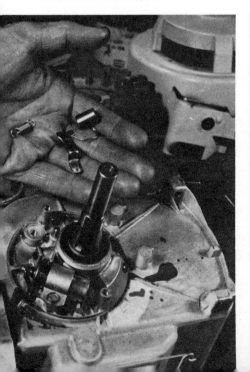

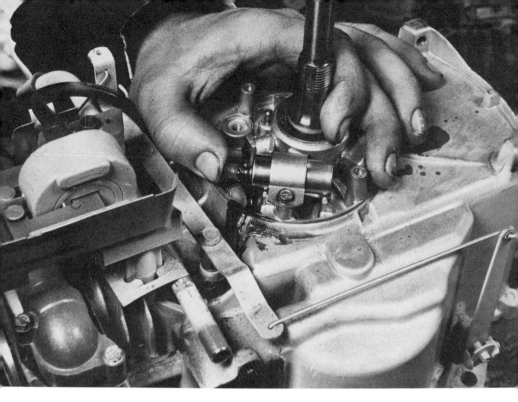

6-31. Disconnecting the primary wire leading to the fixed point is easier if you use the plastic cap that comes with a replacement set of Briggs points. Put the cap on as shown; press in and pull down the wire. Because the fixed point is part of the condenser, the condenser is always replaced with the points on a Briggs engine, an exception among small gas engines.

194

6-32. Checking the point gap with feeler gauge is the final step. Aren't you glad you set the old points for the widest opening before you started? You can't see the lobe on this engine. Only the plunger is visible; and it's in the right position for gapping the new points.

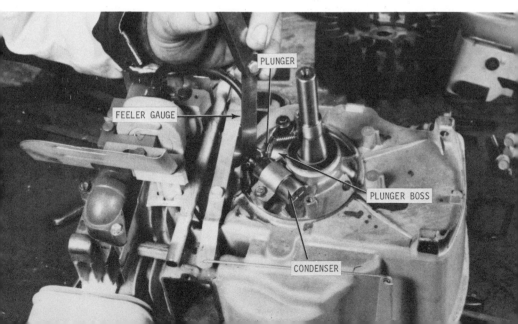

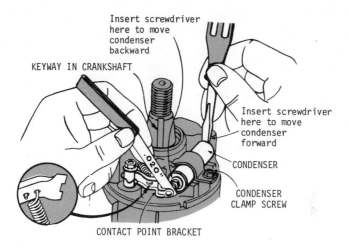

6–33. If adjustment is necessary, just slacken the condenser clamp screw and either push the condenser forward or backward to change the gap.

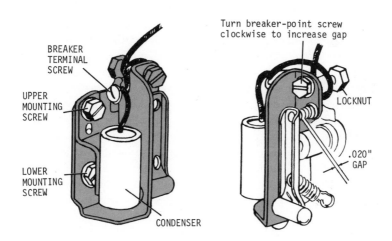

6–34. Here are the points and condenser setup and the adjusting screw arrangement that change the point gap on the Briggs ignition system in 6–43.

use of pliers also does the job. Just make sure the pliers are kept off the fixed point contact face.

Lightly tighten the fixed point lockscrew, then insert the appropriate size feeler gauge between the points and adjust the position of the fixed point as necessary to produce a light-to-moderate drag as you insert and withdraw the feeler. The specifications for point gap may be in the owner's manual. If they aren't, check at the place where you buy the new points. In virtually all cases, the gap should be .020 inch. However, there are exceptions, including McCulloch chain saws (.018 inch) and Sears chain saws (.020 inch for new points, .017 inch for used points). The reason for the variation on Sears saws is interesting. The rubbing block is designed with a softer tip to wear down very quickly, producing a nicely contoured surface against the crank; then the wear virtually stops. On used points, the wear already has taken place, so the reduced gap is the one which the new points will have within a very short time.

Leaving the lockscrew lightly tightened will enable you to change the gap as necessary without having the fixed point plate shift excessively during the maneuver.

BREAKER PLATE AND TIMING

The spark's arrival at the plug is a precisely-timed event, and well it should be; refer back to 6–10. If the spark arrives at the wrong time, the combustion pattern in the cylinder is upset, and the engine runs poorly. Normally, the spark timing does not change significantly in a typical magneto ignition system. The only things that can disturb the setting are these: 1) movement of the magneto coil in a way that changes its relationship to the flywheel magnets; 2) a change in the breaker point gap; 3) loosening (and resulting movement) of the breaker housing. On many small gas engines, adjusting the gap between magneto coil and flywheel will not affect timing; on others, the magneto coil position is the spark timing adjustment. The simple ways to tell:

1. If the breaker points are in a housing or on another plate that is held by bolts in elongated holes, then timing adjustments are made by moving the plate or housing. In this case, the magneto coil may be bolted to a flange that is part of the breaker housing, so that any movement of the housing also repositions the coil relative to the flywheel.

2. If however, there is a raised arrow on the flywheel near the circumference, then the magneto position alone permits adjustment of the timing. If you have to remove the breaker points plate, or move the magneto (to adjust gap to flywheel), scribe realignment marks if the engine doesn't have them. In the case of the flywheel, just line up the arrow with the magneto coil and put a dab of paint or nail polish on the coil directly op-

posite the arrow. If you are replacing the breaker plate, you must retime the engine. Even if there are timing marks on the old plate, they cannot be transferred to the replacement part.

READJUSTING IGNITION TIMING

(Note: Refer to 6–10, 35, 36 and 37.) It is possible for the magneto screws coil or breaker points plate to work loose during operation and disturb the ignition timing. Or in the case of the Briggs engine, a small amount of wear on the point opening plunger can affect the timing by causing the points to open a bit late.

On engines with the timing mark on the flywheel, adjustment is simple: 1) Insert a piece of thin cigarette pack cellophane between the breaker points. 2) Turn the crank until you can just remove the cellophane (no drag); the breaker points have just begun to open. 3) Put the flywheel back on, held in place by the half-moon key. The flywheel arrow should be directly opposite a mark on the magneto coil. If there is no reference mark on the magneto, the manufacturer probably is using one edge of the coil lamination as a reference. On McCulloch, for example, the leading edge is the reference line. Check with the parts supplier. 4) Slacken the magneto coil screws and reposition the coil as necessary. If there is a timing mark on the breaker plate, loosen the mounting bolts and realign. See 6–35.

If the engine has no timing mark, the procedure is somewhat more complex, for the manufacturer probably has specified a position of the piston

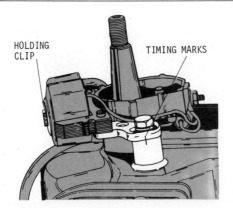

HOLDING CLIP

TIMING MARKS

6–35. This Tecumseh breaker housing allows you to make very precise ignition timing adjustments. The magneto coil sits on the housing, and it's automatically in position when the two timing marks line up. Then to properly time the engine, just set the point gap. If you replace a breaker housing, of course, the ignition system must be timed as described in 6–34; or as in the following ones.

just before it has reached the top of its stroke in the cylinder. What you must do is remove the cylinder head, as described in Chapter 5, and turn the crankshaft (by turning the flywheel) until the top of the piston approaches the top of the cylinder. On two-stroke and four-stroke engines with crankshaft-operated points, the spark occurs every time the piston is at the top of the cylinder. (On four-strokers, the spark that occurs when the piston is on its exhaust stroke is unnecessary, but harmless.) On four-stroke engines with camshaft-operated points, the spark arrives only when the piston is coming up during the compression stroke. You can tell compression from exhaust with the cylinder head off because both valves are closed during compression. And on these engines, you set timing when the piston is making the compression stroke.

The factory specifies the distance before the top of the cylinder in two ways: 1) degrees before top dead center and 2) fraction of an inch, or actual distance, as measured with a small ruler, that the top of the piston is below the top of the cylinder. Unfortunately, the dimensions are usually given in decimal fractions, so you must convert to the closest common fraction. Example: All Tecumseh four-cycle engines below 11 cubic inches

6–36. If the ignition timing has been disturbed and there are no alignment marks scribed, remove the cylinder head and bring the piston up on the compression stroke, stopping a specified distance before top dead center. On this particular engine, a Tecumseh, the distance is ¹⁄₁₆ inch and is measured with a machinist's ruler.

displacement have a .060 inch specification. All those above 11 cubic inches, .090 inch, except the 9.06-cubic-inch engine with a horizontal crank, which is .030 inch. The .030 inch is set for 1/32, the .060 for 1/16, and the .090 for 3/32 inch.

The ruler you use should be the finely-marked machinist's type. If you're going out to buy the feeler gauges, look for a set that includes such a ruler. See 6–36.

Convert Timing in Degrees

If timing is only given in degrees, you can convert that figure into a decimal and then into a common fraction. For example, timing is specified as 26 degrees before top dead center on an engine with a piston stroke of 1.495 inch. To convert, multiply 1.495 by 26 and divide by 360, which gives the result of .108 inch. The fraction 7/64 inch is .109 inch, so set the top of the piston at 7/64 inch before the top of the cylinder. Note: This method is sufficiently accurate and should be used only if timing is in the high teens to high twenties. More precise conversions involve advanced math.

6–37. Next, slacken the breaker housing bolts and reposition the housing as necessary so that the breaker points are just opening. (Since it's hard to see the point opening by just looking at the parts, first place a piece of cellophane between the points, and slip it out as the points crack apart.)

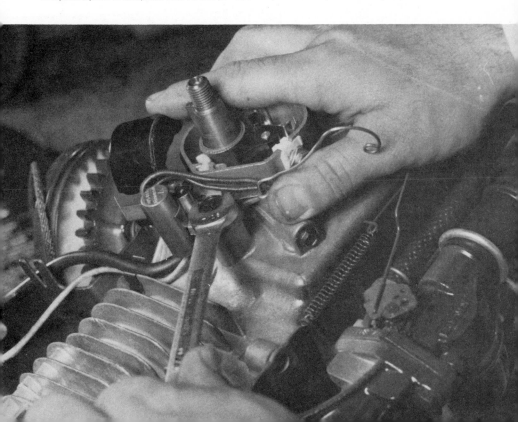

Making the Adjustment

Once the piston is in the proper position, the breaker points should just be opening. You can check this with a piece of thin cellophane. (You should be able to just slip the cellophane out, without drag, when the points just crack apart.) If the points still are closed when the piston is in the correct position, and the breaker point adjustment is correct, loosen the breaker plate bolts (6–37) and shift the plate so that the points just open. A simple way to do this is to shift the plate back and forth to see which direction causes the points to open, and which to close them. Shift the plate to close the points, insert the cellophane, then slowly move the plate in the other direction until the cellophane is just free. Now to double check the setting, tighten the breaker plate bolts and turn the crank until the points close.

Continue turning the crank until the points just open (cellophane free). Measure from top-of-the-piston to top-of-cylinder, and the dimension should correspond with specifications.

REPLACING THE CYLINDER HEAD AND MAKING A SPECIAL TOOL

Whenever you remove the cylinder head, you must install a new cylinder head gasket; this is covered in illustrations 6–38 through 42. Although

NEW CYLINDER HEAD GASKET

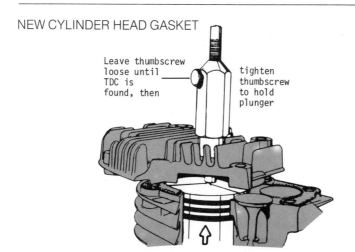

Leave thumbscrew loose until TDC is found, then — tighten thumbscrew to hold plunger

6–38. This special tool makes it unnecessary to remove the cylinder head a second time to measure the distance from the piston to the top of the cylinder. The tool threads into the spark plug hole. The plunger can be locked with the thumbscrew once you've brought the piston up to top dead center, something easy to see when the cylinder head is off. Install the head and gasket.

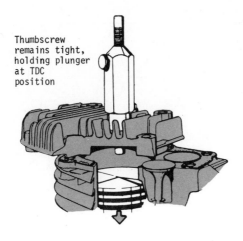

Thumbscrew
remains tight,
holding plunger
at TDC
position

6-39. Turn the flywheel counterclockwise to bring the piston down slightly, and then check the timing specifications.

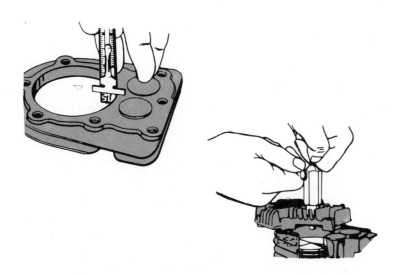

6-40. Each mark on the plunger is 1/32 inch, permitting you to determine how many marks the plunger must go down. Slacken the thumbscrew. Move the plunger down the appropriate number of marks. Then tighten the thumbscrew to lock the plunger.

head gaskets for small gas engines aren't expensive (a few dollars), you probably won't want to install a new one every time you check the timing. You can buy a special gauge that fits through the spark plug hole to measure the distance of the piston below the cylinder. Or you can follow this less expensive procedure: Simply mark the flywheel and a metal part of the engine directly across from your flywheel mark. It doesn't matter where you make the lineup, so long as you don't mark the magneto coil, which you might replace some day. A hammer and pointed punch will do nicely, and then you'll have permanent markings. It's possible you might have to replace the flywheel, but if the flywheel is keyed into position, the marking can always be transferred to the new one.

On automobiles, a camshaft operates the valves and the breaker points. If you have one of the rare small gas engines whose points are operated by a camshaft, you'll find the points mounted externally, as on the automobile. Thus it's not necessary to remove the flywheel to replace the points. See 6–43.

OTHER IGNITION MAINTENANCE

An ignition system inspection should always include careful scrutiny of the magneto coil laminations (6–44) and all wiring. If any wiring has bared insulation, this can cause short circuits and loss of ignition. Cover frayed insulation with plastic electrical tape, being careful not to overdo a good thing. If you put on too much tape, you might, for example, interfere with

6–41. Any time you want to check the timing, just insert the tool and bring the piston up until it is in contact with the tool plunger. (On a two-cycle chain saw engine, this tool can be easily used without taking anything apart. You can see the piston through the exhaust port when you have removed the muffler, so you can easily tell when it's stopped moving up.)

Be sure screw is tightened so piston won't disturb BTDC position

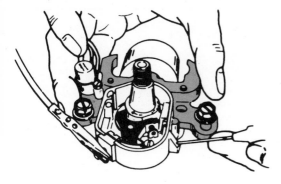

6–42. The final step of the timing job is to slacken the breaker housing bolts and position the housing so that the points are just opening.

EXTERNAL BREAKER SETUP

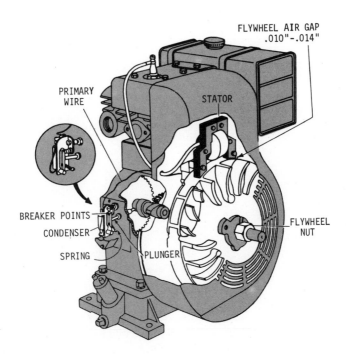

FLYWHEEL AIR GAP
.010"-.014"

PRIMARY
WIRE

STATOR

BREAKER POINTS

CONDENSER

SPRING

PLUNGER

FLYWHEEL
NUT

6–43. There are some small gas engines in which the entire ignition system is outside the flywheel, such as with this Briggs setup.

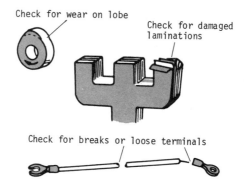

Check for wear on lobe

Check for damaged laminations

Check for breaks or loose terminals

6-44. Not all ignition system problems are caused by worn breaker points, a defective condenser, or flywheel defects. Also check 1) the wire from the points to the magneto coil for breaks or looseness at the terminals, 2) the magneto coil assembly for damaged laminations, and 3) if the points cannot be adjusted to close completely, the crankshaft lobe itself. The lobe should open the points to a specified gap and on some engines is a replaceable part. You can check it by unbolting and lifting off the breaker housing (after scribing alignment marks).

the fitting of the breaker points cover or the action of the points themselves.

If the magneto coil laminations have separated, replace the part. Don't fool around with trying to glue them back together.

Oil on the Breaker Plate

An oil film on the points and breaker plate is a common problem on Briggs and Stratton engines, caused by leakage past the plunger that pushes the points open. Briggs has a special gauge that can be used to check the plunger hole, and any Briggs shop should be able to make the check in a few seconds if you bring in the engine with flywheel and points removed. If the gauge goes in more than a quarter inch, the plunger hole is excessively worn and a new circular sleeve, called a bushing, must be installed. Inasmuch as this job requires removal of the crankshaft and a special reaming tool, it's best left to the repair shop.

There are other possibilities that you should check before passing off the job: 1) The plunger may have been installed incorrectly; the grooved end should be against the movable point. 2) Oil may be seeping past the breaker plate holes for the points' lock screws; in this case, coat the screws with a nonhardening automotive sealer, such as Permatex, before installing.

On badly leaking Briggs engines with the points operated by the camshaft, there is a plunger seal kit available. The kit has been installed in factory production for several years, but if your engine is an oldie, it may not have the seal, or the seal may have been damaged in installation (something very easy to do). Check with the parts supplier to obtain a kit if your engine was built before 1970, when the production change was made.

Note: Defective crankshaft seals on any small gas engine also may result in an oil film on the breaker plate.

Dust in the Breaker Housing

If dust gets into the breaker point housing, it can accelerate wear on the fiber block and the lobe or the plunger. This is a common problem on chain saws, but it also can occur on mowers. Always clean the housing, using a vacuum cleaner if necessary. Then coat the edge of the breaker point cover (and any opening through which the wiring passes) with silicone rubber sealer.

Starters

A starter is a device that spins the flywheel, which is attached to the crankshaft, setting in motion the pistons and rod, ignition and fuel systems to get the engine running on its own. In an automobile, there's only one design in modern use: an electric motor supplied with current from a storage battery. You'll find this on the large riding mowers and snow blowers, but the basic little machine you have is probably equipped with a somewhat less expensive and much smaller arrangement.

The most popular design is the pull-rope spring starter. You slowly pull out the cord for the first few inches, then quickly the rest of the way. The slow initial pull spins the flywheel fast enough for the starter to engage the flywheel by one of various pieces of hardware in wide use. The rapid pulling spins the flywheel. If the engine doesn't start, you release the cord and a spring rewinds the drum around which it's wrapped.

Some small engines employ a wind-up starter. You turn a handle that is somewhat like a ratchet wrench. (In fact, it really is a ratchet wrench that turns a starter drum to wind up a spring.) You pull a release lever and the spring unwinds a clutch device in it, engaging the flywheel and spinning it.

The electric starter is available in three varieties. The smallest is a type you insert into the engine and plug into household current, and it then spins the flywheel. When the engine starts, a clutch device kicks the starter free of the flywheel.

Another design is built into the engine, but plugged into household current. It is very similar to the battery operated model, except for the absence of the battery and a special switch.

PULL-ROPE STARTER

This book devotes most of its attention to the simple pull-rope design, because the odds are overwhelming that this unit is what you've got. Refer to 6–45 through 65.

The rope starter looks so simple you would think nothing could go wrong. Unfortunately, the following are typical problems:

1. The rope breaks or stretches.
2. The spring breaks or gets weak.
3. The method of engaging the starter to the flywheel develops problems, and because there are several ways of handling the engagement, there are many possible problems that can develop.

There's one nice thing about rope-starter problems; you can generally figure out what's wrong before you even take anything apart. If the rope breaks or the spring doesn't rewind, that's pretty obvious. If pulling the rope doesn't result in the engine turning, you can safely assume that there is a malfunction in the method of engaging the starter and the flywheel.

The typical rope starter is easy to remove for inspection. It normally is mounted in a top or side cover directly across from the flywheel, and in some cases is part of a separate piece of sheet metal that can be removed without taking off the entire cover. Whatever the situation, getting to the starter is not a problem. Some rope starters are behind the flywheel, which means that the flywheel must come off first. Once it's off, however, you'll find this starter basically the same as any other rope type.

Working on a starter, however, demands some respect, because you're dealing with a pretty strong spring that could hurt you or others if it were casually handled and allowed to unwind without restraint. The homeowner normally does not have the torch equipment necessary to heat a spring until it is desensitized. An important thing to remember is that the spring is not slack even if the starter cord has been completely rewound, or even if the spring seemingly unwound when the starter cord broke.

If the spring is intact, but it doesn't seem to be working properly, possibly because of a cracked drum or housing, you must release as much tension as possible before trying to disassemble the starter. (It isn't just ordi-

6–45. To service a typical rope-type rewind starter, you begin by removing the mounting screws.

6–46 and 47. As shown below left, pry off the plastic pilot sleeve. Then use a pliers to pull the rod (that holds the pilot sleeve) out of the retaining screw.

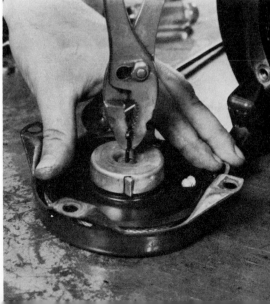

6-49. Once the screw is loose, it can be removed with an ordinary Phillips screwdriver.

6-50. The screw and cap come off.

6-48. A hammer-actuated hand impact tool with a Phillips head loosens the lockscrew.

LOCKING TANG

PULLEY

SPRING COVER

END OF SPRING HOOKS ON HERE

6–51. Next step is to lift the pulley up and out, exposing the spring cover, which is held by tangs that lock to the pulley body.

6–52 and 53. Twist the spring cover clockwise to disengage it from the pulley; then lift it away, as shown left. For this model, you can obtain a replacement cover with the spring installed. To install a new rope, just run the old one out through the hole, first digging it out with a small screwdriver.

6–54. Knot the new rope at the end, and seal the knot with a match.

6–55. Starter dogs should be checked to see if they have good spring tension. If not, lift them out and install new springs.

nary clothesline.) On most starters you can pull the rope off the drum pulley, reaching in with needle-nose pliers, to allow the spring to unwind some more.

On many starters there is a notch in the drum pulley, which will make safe spring unwinding a bit easier. Hold the drum with your thumb, work out a section of cord from the drum pulley, then hook it into the notch; see 6–61. Slowly release thumb pressure on the drum and it will unwind. Repeat this procedure until the spring is as unwound as it will become.

Because of the variety of ways the rope starter is assembled, you have to look at which you've got and see what holds what. The examples given in the rest of this chapter should fit your starter closely enough to enable you to service it.

Pay particular attention to the rope at the pulley end. It may be held in place by a knot, or by a pin through the knot (6–62), or by a pin through the fibers. The rope may go inside a guide or outside; take careful note so you can reassemble correctly (6–63). The pin position is important, or if there is a knot, the type of knot is crucial. A replacement rope should be approximately the same length and exactly the same thickness as the original. And if the ends are unfinished, avoid the possibility of problems from unraveling by holding a lighted match to the ends. (The fibers will melt together. Allow them to cool thoroughly before using the starter.) You should also seal the knot by applying the heat of a match for a few seconds.

If you have difficulty guiding a new rope into place, wrap a piece of stiff wire (such as piano wire) around it, and bend the wire as required so it moves into the pulley at the correct angle.

6–56. This starter also is easy to service. The cover is held by two screws.

6–57 and 58. Lift the cover off, left, and there's the spring. Then if it's necessary to install the new spring, here's how. Pull out the old one. Carefully hold the new spring in its retainer up to the pulley and push the coils into place. This way there's no tedious unwinding and recoiling necessary on the new spring.

6–59 and 60. This chain saw starter spring is the type you have to unwind and recoil, but the job isn't difficult once you have the inner end hook on and a few coils stuffed in. Then you install the starter drum and turn it. This will wind the spring in the rest of the way.

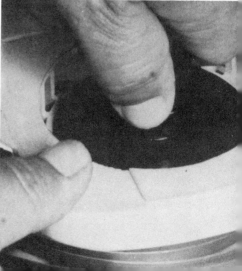

6–61. Installing the rope and setting up proper initial tension are easier if there is a notch in the pulley, as shown.

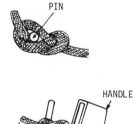

6–62. If the rope is held in the handle with a pin, use a figure-eight knot. Then seal with a flame.

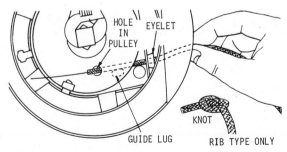

6–63. Watch where the rope goes. Note that it goes through an eyelet first and then inside the guide lug on this Briggs starter. A simple knot is used. To hold tension on the pulley (more than 13 turns), insert a piece of ¾-inch-square hardwood into the pulley hub and turn it with an adjustable wrench. (The guide lug was eliminated on later model pulleys.)

STARTER SPRINGS

The spring normally sits in a housing, which may be part of the engine cover you removed. It is covered by the pulley drum, which is held to a pulley hub by a lock screw or a C-shaped ring. Be very careful when removing this screw; and be doubly careful when pulling off the drum, or the spring may come flying out. In most cases you can disengage the spring from the drum by turning it in a direction opposite to engine rotation and lifting gently. Keep a screwdriver in hand to slip underneath the drum to hold down the spring, just in case.

The replacement of a spring may be as simple as installing a new housing with spring wound in (6–52), or as difficult as winding in the new spring, loop by loop. If the spring tension is not excessive, you can push the spring off the retainer into place (6–58). Start with the outer circumference of the spring, and make sure you push the end into the retaining groove. If you're careful, you may be able to get the rest of the spring in without its unwinding. If not, carefully let it unwind and work it in loop by loop. Many small gas engines are designed for this type of operation (6–59). In any case, it's not a terribly difficult job if you remember to observe the way the old spring is installed and make sure you don't install the new one backwards. (It's been done.)

You've installed the new spring and refitted the drum. Now you're ready

---

SPRING TENSION

6–64 and 65. Proper tension on the spring is important. Note that the spring is absolutely tight in this cutaway photograph and that the rope isn't pulled out all the way. In this case, the spring would snap if an energetic effort at starting were made. This spring, as the cutaway shows, is properly tensioned. The cord is pulled out all the way, and you can see some space between the coils, meaning the spring could be tensioned a bit more without breaking. Yet, it will retract the rope smartly.

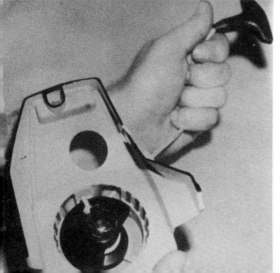
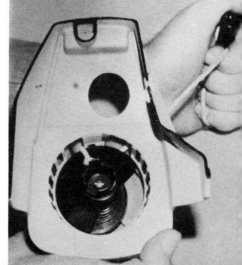

to install the rope, which may prove to be the most difficult part of the job. On most starters the only way to install the rope is through the holes provided in the cover and pulley. In these designs you have to prewind the pulley by hand until it is tight and until the holes are in reasonable alignment; hold the pulley and have a helper slip the cord through the holes and knot it. Then you slowly release the pulley and let the spring unwind, which will wind up the rope all the way. Tecumseh starters are prewound six turns, McCulloch two to three and Briggs a difficult 13¼. In order to achieve the heavy tension called for on the Briggs starter, you must obtain a 4-inch square piece of ¾-inch hardwood, and insert it into the drum pulley hub. Put an open-end crescent wrench around the piece of wood and turn the wrench. When you get to the 13¼ turns, the hub and housing rope holes will line up. Get a friend to help you with this one. See 6–63.

On some starters, fortunately, this isn't necessary, because there is a way to wind the drum pulley when the starter is assembled. The notch in the pulley provides a holding point for the rope, permitting you to grab the rope and wind the drum the necessary turns to increase spring tension. Then you release the rope from the notch and the spring unwinds, drawing in the rope.

FLYWHEEL ENGAGING PARTS

There are many ways the starter can engage the flywheel. The object of all of them is to lock the starter to the flywheel during your operation of the pull starter, then to disengage the starter from the flywheel when the engine starts.

The Briggs system consists of a type of clutch. As you pull the rope, you spin a propeller-like part that fits inside a sealed housing with steel balls in grooves. See 5–64 and 65 in Chapter 5. The housing is so shaped that when the propeller-like part is turned by pulling the rope, it locks against the balls in the grooves. When the engine starts, you stop pulling the rope, and the balls are spun around the propeller instead of locking to it.

Another system features hook-like parts called pawls which are mounted on spring-loaded pivots on the flywheel exterior. Projections from the starter engage the pawls, and the flywheel spins when you pull the starter rope. When the engine starts, centrifugal force spins the pawls away from the starter projections. When the engine stops, the springs return the pawls to the start position.

The dog clutch is still another popular design. The dog is a rectangular piece of metal mounted inside a slot in a cylindrical drum and is kept away from the flywheel by a spring (6–55). When you pull the starter rope and thus spin the drum, centrifugal force pushes the dog out against the projection on the flywheel, where it locks, pulling the flywheel around with the drum. When the engine starts, you let the rope go and the dog springs back, out of contact with the flywheel.

Some rope starting systems are actually very similar to the electric starters used in cars. If you look at a car's starter and flywheel, you will see a large gear around the circumference of the flywheel and a small gear on the end of the starter. When you turn the key, the starter motor spins, and the turning of the motor shaft spins the little gear forward on curved splines, into mesh with the flywheel gear. When the engine starts, you let the key go. The starter shaft stops and a spring retracts the little gear away from the flywheel. The rope system simply substitutes your pull on the rope for electricity spinning the starter motor shaft.

The parts that lock the flywheel to the starter during the rope pull are usually reliable. On the pawl or dog type, simply make sure the springs have reasonable tension and are properly hooked on. See 6–55.

The Briggs clutch eventually wears out, and although it can be disassembled by prying the halves apart, the part is cheap enough to make replacement advisable. The clutch may, however, come apart during removal, in which case it's worth fixing. Just coat the balls with grease, fit them into the outer recesses of the groove, put the housing halves together and crimp back together with hammer and punch marks around the circumference. See 5-64 and 65 in Chapter 5.

The gear setup may suffer from chipped teeth, a broken starter gear return spring, and sticking of the gear on its curved splines. If the flywheel gear breaks, installing a replacement is a job for the machine shop. The starter gear and spring are usually held by a snap-ring arrangement and can be replaced without difficulty. To avoid gear sticking, spray penetrating oil on the splines once a season.

WINDUP AND ELECTRIC STARTERS

The windup starter has a very strong spring and its service, aside from removal for engine service, is best left to the professional. If you insist on trying, however, you'll find this starter basically similar to the rope starter except that a ratchet handle is built into the top and there is a locking dog that holds the tension developed by turning the handle. There is also a release lever to pull the locking dog away, allowing the spring to spin the flywheel.

Electric starters can only be checked with special equipment. The only normal maintenance is an application of penetrating oil to the gear splines. (Starter removal procedures and application of penetrating oil are shown in 6–66 through 73.)

On a machine with a storage battery, check the cables for tightness, and also check the battery water level. If the battery posts and cable terminals (inside and outside) are coated with white corrosive deposits, clean them with a wire brush.

6-66. If your mower has an electric starter, don't let it intimidate you. This type plugs into household electric outlets, and as you can see, it's held by some Phillips-head screws.

6-67 and 68. After the screws are out, just lift the starter away from the engine. Then oil the starter on the bench. Turn the small gear to expose the splines, the only area needing periodic lubrication.

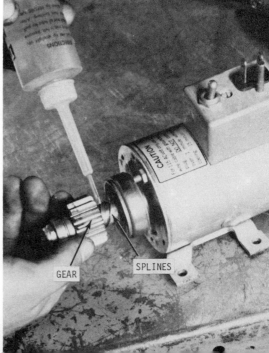

GEAR

SPLINES

6–69. If you have a riding mower, begin removal of the electric starter by disconnecting the battery cable that is bolted to the chassis (the ground cable). This immobilizes the electrical system to insure that you don't get a shock or burn out a part by causing an inadvertent short circuit.

6–70. The engine top cover must come off on this model to provide access to one of the starter mounting bolts.

STARTER GEAR

6—71. Disconnect electrical terminals at the top of the starter if you are removing it for service or replacement. If you just want to oil the splines, you can do it with the starter in the chassis; just turn the starter gear to expose them.

6—72 and 73. To replace the starter, remove the mounting bolts. The bolt in the left-hand photo was hidden by the engine top cover. With the starter out, disconnect the cable from the battery.

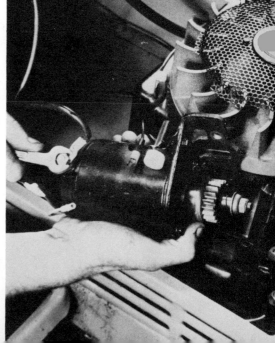

7

Servicing Fuel Systems

A wise man once said, "The best service you can do for a small gas engine fuel system is nothing." He was exaggerating a bit, but it's probably true that more harm than good is done by tinkering with the heart of the system, the carburetor. If the engine started and ran acceptably the last time you used it, that is a very strong indication that the carburetor adjustments are correct. So at least observe this rule: Keep your hands (and screwdriver) off the blankety-blank adjustment screws unless you know exactly what you're doing.

The two most important fuel system services you can perform do not even require that you know the inside of a carburetor from the windings of your central nervous system. If you perform them faithfully, you'll avoid a lot of carburetor and engine grief and probably save your central nervous system a lot of aggravation.

FILTERS

Number 1 is regular service of the air cleaner, which has the job of filtering the air before it goes into the carburetor air horn. If the filter is clogged, it restricts the air flow, sharply reducing the performance of the engine. If you just pull off the filter, dust and dirt will wear out the engine in an unhappily short time.

Most filters on small gas engines are either the replaceable paper element type or a cleanable plastic foam, and occasionally you find them made of a metal mesh. The filter is connected to the carburetor air horn, either directly mounted or connected by a hose. See 7–1 through 7–4.

The paper type is of a pleated design that although built for replace-

AIR FILTERS

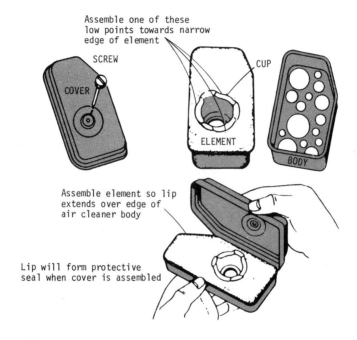

Assemble one of these
low points towards narrow
edge of element

SCREW

CUP

COVER

ELEMENT

BODY

Assemble element so lip
extends over edge of
air cleaner body

Lip will form protective
seal when cover is assembled

7–1. You saw this Briggs and Stratton polyurethane foam air filter in Chapter 5. Here it is again, disassembled for cleaning and oiling.

ment only can be cleaned to some extent. Just tap it against a piece of wood to dislodge loose dirt from the exterior; then vacuum clean. You won't get all the dirt out, but you'll remove enough to extend the element's life, perhaps for as long as two seasons. A dirty paper element usually looks dirty, but if sand is in the air, it might not show on the filter. As a quick test, run the engine for about half a minute without the filter, and if performance is markedly better, install a new element.

The plastic foam filter is cleaned in solvent. Remove it from its support cage. Then wash it in kerosene, automotive solvent such as Gumout, or low-suds household detergent and water. Wrap it in cloth and gently squeeze out the cleaning agent until the foam is "damp dry." Fill the foam with clean engine oil (SAE 30), then gently squeeze to remove excess oil. See 7–2.

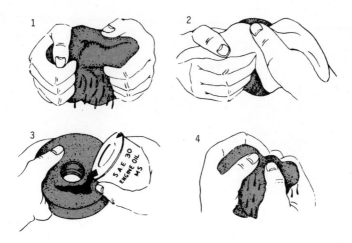

7-2. Here are the four steps of polyurethane-foam filter service: 1) Wash the foam in kerosene, carburetor solvent or liquid household detergent and water to remove dirt. 2) Wrap the foam in cloth and gently squeeze dry. 3) Saturate the foam with clean SAE 30 engine oil. 4) Squeeze gently to remove any excess oil.

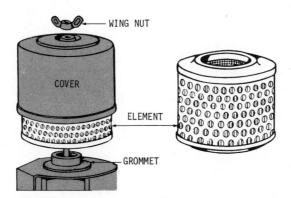

7-3. A pleated paper air filter is partly cleaned by tapping the element (top or bottom) on a flat surface. When the element is dirty and tapping doesn't seem to remove much dirt, install a new element.

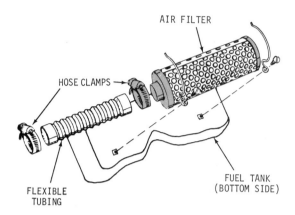

AIR FILTER

HOSE CLAMPS

FLEXIBLE TUBING

FUEL TANK (BOTTOM SIDE)

7—4. This paper element is not mounted directly on the carburetor. Instead it is held to the underside of the gas tank by a clip and is connected to the carburetor by a hose. To prevent unfiltered air from reaching the engine, be sure to check the hose clamps for tightness at both ends.

If you happen to have one of those rare metal mesh filters, clean in automotive solvent, or a low-suds detergent and water solution. Allow it to air dry.

GAS TANK

Emptying the gas tank at the end of every season prevents the formation of varnish that results from the mixture of gasoline with air and moisture. This is a simple, effective step that extends the life of the engine and the carburetor.

CARBURETION

Removing the carburetor is generally an easy step on small gas engines, and the typical procedures are covered in Chapter 5. Rebuilding the carburetor with a kit also boils down to a series of reasonably simple steps, covered later in this chapter. It is not absolutely necessary to understand exactly how a carburetor does its job in order to be able to overhaul it, because the manufacturer thoughtfully includes all normally needed replacement parts. All you have to do is take the carburetor apart, clean it in solvent, and install the new parts.

If you have studied troubleshooting in Chapter 4, you will know when the fuel system is causing problems. Then the well-under-$10 investment in a carburetor repair kit won't be wasted. In general when a small gas engine is a few years old, properly done major carburetor service is beneficial. Note the words "properly done"; they're there for a reason. The small gas engine carburetor is a small component, and although working on it hardly requires the craftsmanship of an Old World jeweler, the thing is not something you just slap around.

Most people couldn't care less about the gory details of carburetion theory. Then, too, a little knowledge can be a dangerous thing. But in this book, we attempt to strike a balance, giving you the minimum amount of theory and a reasonable amount of how-it-works. This should enable you to approach the carburetor with some degree of confidence and, if necessary, help you figure out the little nuances incorporated in every design by the different manufacturers.

Chapter 1 provided a basic introduction and if you thought that was enough, proceed to the section of this chapter headed "External Carburetor Adjustments" and continue from there. But if you would like an added dose of basic operation, here it is.

Let's begin with the simplest carburetor and build it up from there. Probably the most basic design in current use is shown in 7–5, and you'll find it on some Briggs engines. This is the suction lift carburetor, and it's always mounted on top of the gas tank, with a pipe in its base projecting into the tank. The tube has a restricted section which is covered by a ball, called a ball-check valve (7–6).

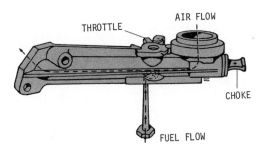

7–5. This is a basic small gas engine carburetor. Fuel is sucked up the tube and through the passage in the carburetor. Air rushing in through the air horn draws fuel droplets out of the passage and carries them into the engine.

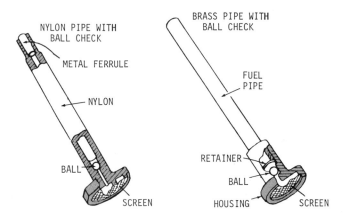

7–6. These are fuel pipes with ball-type check valves.

The air rushing through the air horn section of the carburetor into the cylinder creates the vacuum (like the perfume atomizer) that pulls the ball up and allows the fuel to be drawn up the pipe and into a carburetor fuel passage, called the jet, then through the jet into the air horn, where it mixes with the air.

The fuel flow in the jet is controlled by a tapered needle threaded into it. The farther the needle projects into the passage, the less fuel that can flow through. If the needle is turned out, the passage is less restricted and more fuel can flow through. This needle is called the needle valve, and it is adjustable by turning the screw head on the exterior. A further limitation on fuel flow is provided by carefully sized metering holes in the jet.

The throttle plate pivots in the air horn to regulate the air flow. You control this plate with a trigger on a chain saw, a lever on a mower or blower, the gas pedal on a car.

A second plate, the choke, provides a second means of regulating air flow. It is located in the air horn before the throttle, so that its position exerts an effect on the fuel mixture by reducing the volume of air that can pass. The choke normally is operator-controlled by turning a knob or pulling a lever.

A slightly more sophisticated design combines a diaphragm fuel pump and long and short pickup pipes. See 7–7 through 11. The short pipe supplies fuel, from a cup section at the top of the tank, to the carburetor—a short, easy trip for better fuel supply during starting. The long pipe proj-

7–7. Let's take an "in-the-metal" look at the twin-pipe Briggs carburetor, removed with the gas tank in Chapter 5. Separating it from the gas tank begins with removal of the two outside screws.

7–8 and 9. The bracket screws come off next. Then lift the bracket itself away. This provides access to the mounting screw underneath.

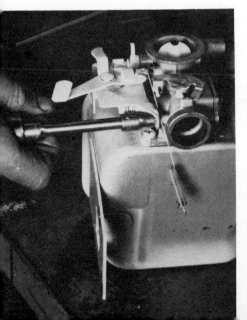

7–10 and 11. Lift the carburetor up and out, as shown left, exposing two pipes and a gasket on the fuel tank. The new gasket should be installed before refitting the carburetor. In the right-hand photo the side cover of the carburetor has been removed, exposing the pump diaphragm (held in hand) and also exposing the gasket.

ects into the main part of the tank, and supplies fuel to the pump section of the carburetor. The pump is the carburetor's main fuel supply system, and it also keeps the cup section filled.

The pump diaphragm (7–11) has inlet and outlet flaps, which serve as flow control valves in response to vacuum pulses in the engine cylinder of a four-cycle, and in the crankcase of a two-cycle. Thus there is no need for a ball-check valve.

Other diaphragm pump systems do not have flaps. They may have fiber balls or discs in specially-shaped passages to control fuel flow. In these cases, the diaphragm action moves the balls or discs.

AUTOMATIC CHOKE

Even such simple carburetor designs as these may be equipped with an automatic choke of sorts. It isn't the temperature-sensitive type used on automobiles. But what do you want in a lawn mower?

A diaphragm under the carburetor is connected to the choke plate shaft

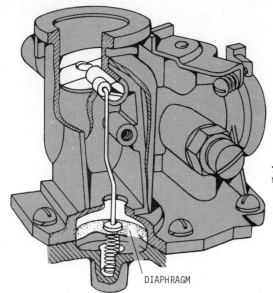

7–12. This spring-loaded choke is automatic, controlled by engine vacuum.

DIAPHRAGM

by a link (7–12). A spring under the diaphragm holds the link down so that the choke is closed when the engine isn't running. Also see 7–14.

Only when the engine starts is there sufficient engine vacuum to pull the diaphragm down (against the spring pressure), opening the choke. This setup also acts somewhat like an acceleration pump on an automobile carburetor. When the engine is operated under heavy loads (cutting thick grass or moving deep snow), engine vacuum drops as the load strains the engine. The drop in vacuum allows the spring to partly close the choke, slightly enriching the fuel mixture to improve low-speed performance.

OTHER DIAPHRAGM CARBURETORS (7–16 through 33)

Most small gas engines have somewhat more precise diaphragm carburetors than the previous one discussed. In these, the diaphragm is not a fuel pumping device. If a fuel pump diaphragm is needed, it is a separate part of the carburetor. See 7–16 and 28.

The diaphragm that controls fuel flow to the air horn is operated in any of these several ways: 1) The diaphragm has a tab in its underside that

7–13. Identifying the jet needle screws and the idle speed screw on a small carburetor is easy. The idle speed screw is the one that doesn't thread into the carburetor body. Instead, it bears against the linkage that moves the throttle plate. The jet needle screws are usually side-by-side, and you distinguish low-speed from high-speed by the size; high-speed is slightly larger in diameter.

7–13a. This simple diaphragm carburetor is similar to the one in 7–13. Note the placement of the jet needle screws and the idle speed screw. Here the low-speed screw is being removed.

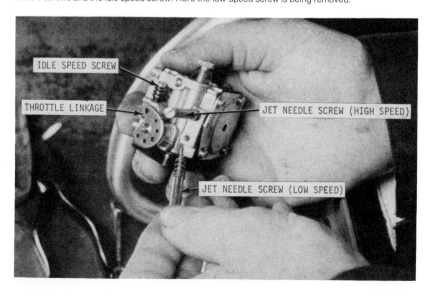

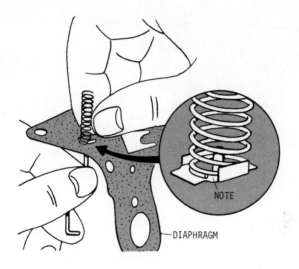

7–14. When servicing the diaphragm, be sure to hook on the choke spring, as shown.

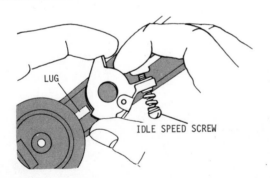

7–15. If you must remove the throttle, look before you attempt it. In the case of this Briggs carburetor, you have to back off the idle speed screw until the throttle will pass a lug when lifted.

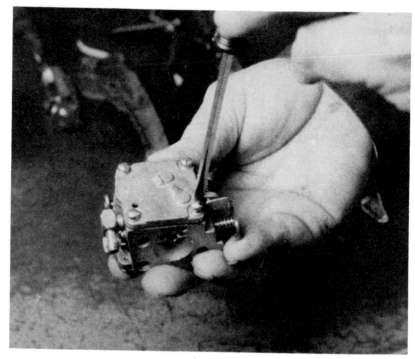

7-16. Remove the cover for the pump section of the carburetor. It's held by the four screws, one at each corner.

7-17. Cover off, you see the gasket and diaphragm. In many cases they stick together and require peeling to separate them.

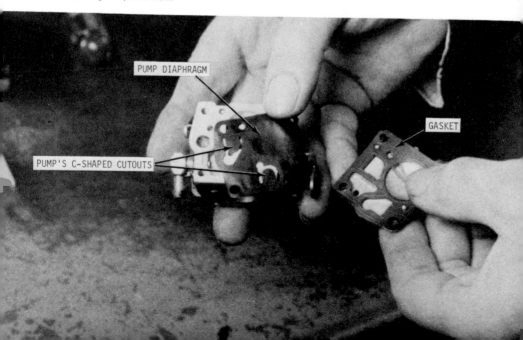

PUMP DIAPHRAGM

GASKET

PUMP'S C-SHAPED CUTOUTS

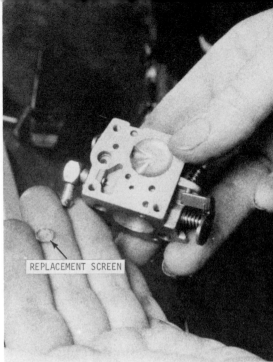

REPLACEMENT SCREEN

7-18 and 19. Carefully pry out the filtering screen with a small-pointed tool, left. The replacement screen in the right-hand photo is supplied in the carburetor repair kit.

7-20 and 21. Remove the carburetor fuel-metering diaphragm cover, as shown in the left photo. Then you can see a notch in the diaphragm rod that retains the rod to the cutout in the fuel-metering needle rocker arm.

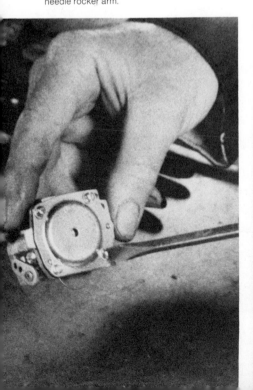

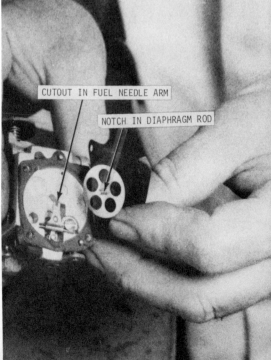

CUTOUT IN FUEL NEEDLE ARM

NOTCH IN DIAPHRAGM ROD

bears against a spring-loaded lever. When engine vacuum draws the diaphragm up, the spring pushes up the lever, lifting a tapered needle that projects into the fuel jet. The greater the vacuum, the greater the diaphragm's effect on the lever and tapered needle. At the point of maximum needle lift, the fuel flow through the jet is greatest. 2) The diaphragm may be connected directly to the lever, and the needle is spring-loaded, so it automatically rises. 3) The needle may be spring-loaded and bear directly against the tab in the center of the diaphragm. 4) The diaphragm action may be reversed, so that the diaphragm moves downward, pushing open the spring-loaded needle or pushing down on the lever. Either of these may be combined with a fuel-priming tube connected to a squeeze bulb much like that used in a perfume atomizer. When you squeeze the bulb, the air under pressure pushes down on the diaphragm, which pushes the spring-loaded needle down. See 7–23 through 7–26.

DIAPHRAGM-NEEDLE CARBURETOR

7–22. This is another type of diaphragm carburetor. Again, the diaphragm has only a tab, but instead of contacting a rocker arm, it makes contact with a spring-loaded needle.

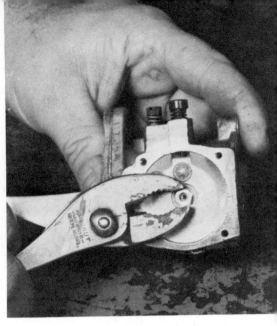

HEX-NUT NEEDLE ASSEMBLY

7–23 and 24. Left, the diaphragm bears against a spring-loaded needle in the center of a hex-nut assembly. At right, you see the wrong way to remove the needle assembly. If the pliers slip, they can damage the hex and possibly bend the needle.

7–25 and 26. You'll need a socket, left, to do the job correctly. In the right-hand photo, the needle valve and spring assembly is being removed.

TWO-DIAPHRAGM CARBURETOR

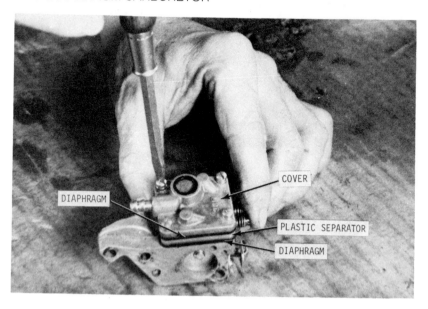

7-27. This McCulloch two-diaphragm carburetor has both diaphragms on the same side, in a sandwich layout.

7-28. The cover is off, exposing the diaphragm (with C-shaped cutout) and the gasket.

7—29 and 30. After removing the white plastic separator, you can lift off the second diaphragm, the fuel metering diaphragm. This one just bears against fuel needle rocker arm. It is not linked to it. Then, as shown right, remove the screw that holds the fuel needle rocker arm.

7—31. Since the needle is linked to the rocker arm, as shown below, lift them out together. Note the little spring which holds the arm up. When crankcase pressure pushes the diaphragm down, the arm is pushed down against the spring. The opposite end of the arm rises, lifting the needle and increasing fuel flow through the carburetor.

7—32 and 33. Above, are the key parts disassembled. Left to right: the needle, the rocker arm and its pivot pin, and the spring. Before reusing, inspect all jet needle screws and fuel flow needles, particularly at the tip. The drawing below shows good and bad needles.

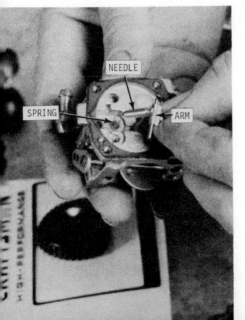

NEEDLE

SPRING — ARM

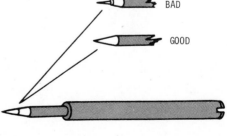

BAD

GOOD

FLOAT-TYPE CARBURETORS

Machinery that stays relatively level usually (but not necessarily) has a float carburetor. This design, which is used in all automobiles that have a carburetor, insures an adequate supply of fuel to the jet regardless of the performance of the pump, which varies according to engine speed and load.

Although the float system resembles that used in a flush toilet, there are noticeable differences. See 7–37, 38, 42, 43. The most important difference is fuel level, which like water level in a toilet, is regulated by the position of the float. To adjust water level in the toilet, you just bend the float rod; and so long as you don't go to an extreme, you'll have at least as much water as needed, and maybe a bit too much. In any case, there won't be much harm either way. In the carburetor, too low a level results in fuel starvation of the engine. Too high a level allows fuel to almost spill out of the jet into the air horn. This excess fuel is not atomized (broken into tiny droplets in the air) and therefore cannot be burned. Severe flooding then causes the engine to stall. Careful adjustment of the carburetor float, therefore, is extremely important.

7–34. Now let's remove a float-type carburetor from this lawn mower engine and see how it works. The carburetor is being removed from the engine together with the intake manifold and the air cleaner assembly. As shown, two Phillips-head screws hold the manifold, which is the air-fuel delivery tube.

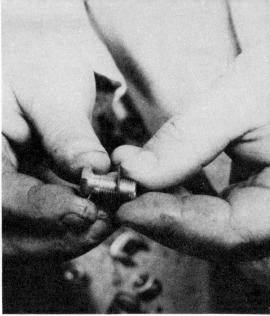

7–35 and 36. Remove the bolt that holds the carburetor bowl, as shown left. When the bolt is out, at right, you can see that it also serves as a fuel jet. The tipoff is a tiny pinhole between the threads and the bolt head. Make sure this tiny hole is clear before reinstalling.

7–37 and 38. The doughnut-like part you see, below left, is the bottom of the carburetor float. This device regulates fuel flow. The right-hand photo shows how: Here you see the float being pulled in the direction gravity would pull it were the assembly in its operating position. In operation, as the float drops, it extracts the needle, allowing fuel to enter the bowl. As the fuel is replenished to the desired level, the float rises, forcing the needle to cut off fuel flow.

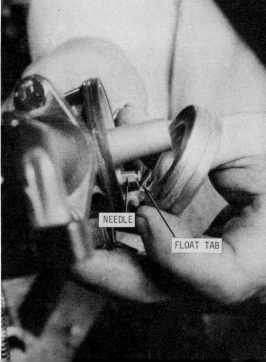

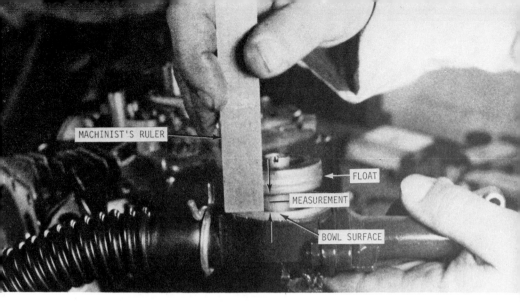

7–39. If you have specifications, you can check float level by inserting a drill bit of specified diameter between the float and the carburetor surface. Without specifications, you can hold the cover or bowl so that the float arm closes the needle valve. Then, with a machinist's ruler, as shown here, measure from the top of the float to the bowl or cover surface at several points around the float. If the measurements differ, bend the float tab to make the float level.

ANOTHER FLOAT-TYPE CARBURETOR

7–40 and 41. Here's another float carburetor, and in this case the bowl retaining nut and bolt setup is obviously a jet. Otherwise, why would the bolt project so far from the nut? Removing the jet exposes a fuel-flow pinhole between the threads. Note the bowl drain, in the photo below right, located to one side of the jet hole.

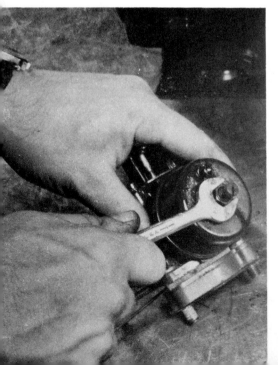
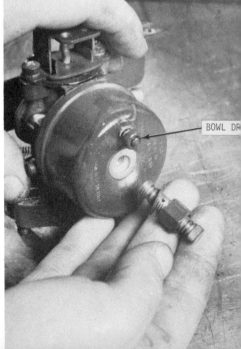

7–42. Remove this float by simply pulling out the pivot pin.

7–43 and 44. Here again, the needle is held to the float by a clip, as shown left. Then use a screwdriver to dig out the filtering screen, which should be replaced as a routine part of overhaul.

MULTIPLE JET ARRANGEMENTS

The basic carburetor described early in this chapter had only one jet. Most small gas engine carburetors have two and sometimes three. The smallest jet provides fuel for low speed and idling, the largest for high speed, and the middle-size for in-between. In actual operation the jets are so positioned that they are in operation according to the amount of air rushing through the air horn. The jets normally work in a progressive order. That is, the medium speed jet (if used) comes in to work with the low-speed jet at medium speed. At high engine speed, the high-speed jet joins the other two. See 7-13 and 13a.

A jet may be a fixed orifice type (just a carefully-sized fuel passage) or adjustable. (A threaded needle valve can be moved in or out to decrease or increase fuel flow.)

EXTERNAL CARBURETOR ADJUSTMENTS

A carburetor normally has only two basic types of external adjustment: the fuel mixture, which is controlled by one or more jet adjustments; and the idle speed, which is regulated by a screw on the throttle linkage. If the engine has been running well, and suddenly fails to start, don't make any adjustments, for the problem undoubtedly is something other than mixture or speed adjustment. Usually, the screws should be touched only if spark plug inspection indicates an overly rich or lean mixture. (Black dry soot on plug electrodes and insulator shows richness; a white blistered look shows leanness.) You should also reset the adjustments whenever you take the carburetor apart for service.

MAJOR CARBURETOR REPAIRS

Servicing a small gas engine carburetor is not difficult if you have the right tools. Some relatively small socket wrenches and screwdrivers are required, for any attempt to force things out with inappropriate tools will only lead to damaged parts. You'll also need some general purpose automotive solvent. Although not absolutely necessary, blowing out the carburetor passages with compressed air does help. So if you wish to add this capability, you'll need the type of hose sold in automotive accessory stores for transferring air pressure from one tire to another. (You just thread it into a spare tire inflated to about 32 psi, then hold the other end into the carburetor in such a way that the hose's valve pin is pushed open.) You can also blow the passages clean with a bicycle pump, if you have a helper to operate the pump, or with air from a compressed air tank or tire-inflating compressor.

TAKING THE CARBURETOR APART

Before disassembling a small carburetor, check the choke and throttle plates in the air horn. Move their shafts back and forth, and see if the shafts and plates move freely. If they do not, drop some solvent into the shafts at the points where they disappear into their carburetor bores; then move the shafts back and forth. If the shafts free up, leave the throttle and choke assemblies alone. If not the parts must come out. Additional service hints are in the illustrated charts at the end of this chapter.

The procedure for removing a throttle or choke varies according to carburetor design. In most cases there is a lockscrew in the center of the plate in the air horn; once that's out, the plate itself comes out. Usually all that's necessary after that is to lift the shaft out of its bore.

The Briggs Choke-A-Matic automatic choke is a sample exception. The choke link must be disconnected from the shaft; refer back to 7–12. Another, the old Briggs Vacu-Jet (a single pipe carburetor) has a throttle shaft that can only be removed after you back off the idle speed screw until the throttle can pass a lug when lifted. See 7–15.

A burr or score on the shaft or plate can often be removed with a very mild abrasive, such as car polish. If the problem is a dirt accumulation in the air horn, pay particular attention to the horn during cleaning, perhaps washing it with a soft toothbrush dipped in solvent.

If the carburetor has a pump section, you'll often find a plate in each end held by several small screws. See photos 7–16 through 7–21, as well as 7–22 through 7–26. Or the diaphragms may be on the same end, separated by a thick plastic plate, 7–27 through 32. Remove the screws and the plates and look at the diaphragms. If they are the least bit stiff, replace them. Carefully inspect the pump diaphragm's inlet and outlet flaps, and replace that diaphragm if the flaps are stiff or if there are even the slightest tears or cracks at their flap edges. Caution: Always remove a diaphragm cover plate slowly; ditto for the diaphragm itself. Otherwise, fiber balls that might be underneath will fall out, leaving you to wonder where they must be reinstalled. Most carburetors don't have the little balls, but why find out your model does—the hard way?

Next, remove the jet needles and inspect the tips with a magnifying glass. If there are erosion circles around the tips, replace them (7–33).

On a diaphragm carburetor, take out the needle valve assembly. On some it comes out when you lift the diaphragm (7–21). Or you may have to remove a lever screw (7–30), or the spring-loaded needle itself. The needle is usually part of an assembly threaded into the carburetor, and there is a hexagonal section for a small socket. Never use a pair of pliers (7–24) or tool other than a socket (7–25). You can inspect this needle valve assembly for erosion circles if you wish, but it's almost invariably part of a rebuilding kit.

On float carburetors (7–34 through 44), remove the hinge pin that holds the float and then the float itself. This gives you access to the needle. The part also is included in rebuild kits, and perhaps so is the seat, which normally can be unthreaded for removal. The seat is occasionally a press fit and must be pulled with a tap and installed with a driving tool. Use a small flashlight to inspect the seat and replace it if it's gouged.

When you've got the obvious parts out, look for the not-so-obvious. The kit you bought may be a clue, for you'll probably find replacements for some things you haven't removed, the most prominent possibilities being fuel filtering screens, fuel pipes, welch plugs, jet, and needles.

Fuel filtering screens (7–18) are pried out with a fine-point object. Welch plugs, which are covers for holes formed to make manufacturing cheaper, leak after a while, so if the kit includes replacements, you can just puncture the old ones with a nail and pry them out. See 7–45 through 47. If you're one of those "don't do it if it isn't absolutely necessary" types, you may wish to leave the welch plugs in place, even though carburetor cleaning will not be as thorough. A sign of welch plug leakage is an engine that shows no response to mixture needle adjustment, so base your decision

---

WELCH PLUGS

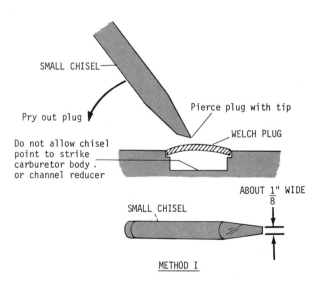

SMALL CHISEL

Pry out plug

Pierce plug with tip

WELCH PLUG

Do not allow chisel
point to strike
carburetor body .
or channel reducer

ABOUT 1" WIDE
8

SMALL CHISEL

METHOD I

7–45. Welch plug removal can be performed in either of two ways. Here you see the most popular way to get the plug out.

on all this information. A coating of nail polish around the circumference of the plugs is a safe sealing procedure whether you install new plugs or not. Fuel pipes come in two types: the nylon one that is threaded into the carburetor base, and the metal one that is pressed in. To remove the metal one, insert the pipe in a vise and pry up on the carburetor as shown in 7–48a. To install a replacement pipe, press it in with carburetor and pipe between the jaws of the vise. The pipe must be pressed in to the same depth as the original was in manufacture, so be sure to measure the length of the projecting section of original pipe before you remove it. These pipes usually have filtering screens in the ends and, on single pipe carburetors,

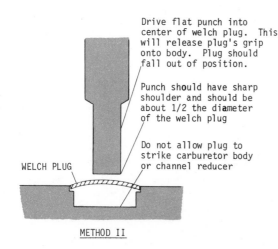

Drive flat punch into center of welch plug. This will release plug's grip onto body. Plug should fall out of position.

Punch should have sharp shoulder and should be about 1/2 the diameter of the welch plug

7-46. This procedure also will work if you don't whack too hard, thereby driving the plug into the carburetor.

Do not allow plug to strike carburetor body or channel reducer

WELCH PLUG

METHOD II

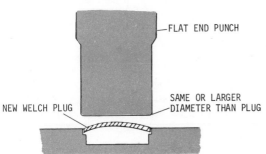

FLAT END PUNCH

7-47. Installing a new welch plug is done as shown. Just flatten the plug. Don't dent it or try to drive it below the top surface of the carburetor. New plugs are included in each carburetor rebuild kit.

SAME OR LARGER DIAMETER THAN PLUG

NEW WELCH PLUG

FUEL PIPES

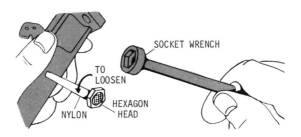

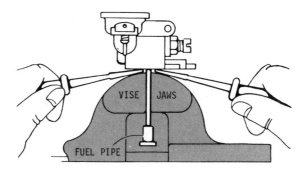

7–48 and 48a. To replace fuel pipes, you should unthread the nylon pipe with a wrench. But you should pry a brass pipe out with the carburetor in a vise, as shown below.

also a ball-check valve. You can shake the pipe and try to hear if the ball is moving, or you can push a thin pin up through the filter screen and feel if the ball moves easily. If it doesn't, clean the pipe in solvent. Ditto if the screen is filled with dirt.

Jet removal is an optional sort of thing. There's no reason why a jet shouldn't come acceptably clean if it stays in the carburetor and the carburetor body is dunked in solvent. If you wish, however, you can remove it to see if it's clogged. There is one notable exception to the "remove it if you wish" rule: the Walbro float carburetor, in which the main jet is cross-drilled at the factory after installation. Once you remove it, there's no way you can install it because you can't realign the cross-drilled holes. Once it's out, you must install a special replacement jet that has an undercut groove, so that the holes no longer need be aligned. See 7–49.

Caution: Nylon parts and diaphragms are not meant to stay in solvent very long. A 15-minute soak, followed by a few minutes of sloshing of the part in the solvent, followed by a five-minute soak and some more sloshing should do the job. So long as no plastic parts remain inside, metal sections can be left in solvent for a few hours and subjected to periodic agitation. Also, if you're going to leave a carburetor in solvent for a while, make sure you really have it apart. If there are synthetic-fiber check balls in some passages that don't pop out when you disassemble, or synthetic rubber O-rings, they could be destroyed by the solvent. So look carefully for them. They may or may not be part of a rebuild kit. When you're buying a carburetor kit, ask the parts salesman if there is anything that might be left in during cleaning that could be damaged. You might discover that the jets are plastic or some such thing.

Check the flat surfaces of the carburetor with a straight edge (such as a machinist's ruler) and a .002-inch feeler gauge. Place the straight edge across the carburetor surface and see if you can slip a .002-inch feeler underneath (7–50). If you can, the surface is warped, and the gasket probably won't seal properly. You could just replace the carburetor, or you can try an additional gasket. In many small carburetors the gasket is part of a diaphragm, so what you'd have to do is get a second diaphragm-gasket assembly and carefully cut out all but the gasket section. Don't jump to conclusions on this, however, for on many carburetors the gasket and diaphragm must be peeled apart (7–17). A second possible cure is to coat

UNDERCUT
ANNULAR
GROOVE

7–49. The main nozzle of the Walbro carburetor cannot be reused if removed because it's impossible to realign cross-drilled holes. The replacement nozzle has an undercut annular (ring) groove, so that proper alignment is automatic.

Service
main nozzle
(reusable)

ORIGINAL
MAIN NOZZLE
Do not reuse if removed

7–50. Before reassembling a carburetor, place a straight edge across the flat surface as shown and try to slip a .002-inch feeler gauge underneath. If the gauge fits under, the surface is warped and the carburetor may have to be replaced; or you might be able to seal the gasket surface, either with non-hardening sealer or two gaskets, or both.

the carburetor surface with a nonhardening gas-resistant sealer, such as Permatex No. 2. This may be the most economical approach, probably the most convenient and just as likely to succeed.

CARBURETOR REASSEMBLY

Carburetor reassembly is basically the reverse of disassembly, but there are some fine points.

When it's available, always buy a carburetor rebuilding kit. It costs about a fourth as much as a new carburetor and will contain only parts that you are likely to replace (7–51). If you have to buy individual parts, get new gaskets and diaphragms, a fuel-flow needle valve assembly, filters and screens.

If you're planning to reuse the mixture needle or lever that bears against the diaphragm, check it for wear. If the tip of the needle is worn, discard the needle. Ditto for the lever if either side shows any wear.

A float requires two checks: 1) Shake it to determine if fuel has leaked in. If you hear the sound of liquid inside, replace it. 2) Inspect the tab that bears against the needle and if it's worn, double check the float level.

On a diaphragm carburetor with the diaphragm hooked to the needle or lever, make sure you've got it hooked back on. In some cases the diaphragm may come off the lever or needle so easily during disassembly that you won't realize it was hooked on. You can tell just by looking at the parts; if the diaphragm has anything but a flat tab in the center underside, it hooks onto something. If the lever top is anything but basically flat and rectangular, such as a C-shape, there's a reason. The C-shape, for example, engages the shank of a very tiny pedestal in the underside of the diaphragm.

After installing the needle and (on those so designed) the lever, place a straight edge across the carburetor surface in a position that also covers the lever or needle. On those carburetors with a gasket separate from the diaphragm, install the gasket first. The top of the needle or lever is normally just level with the surface or perhaps a thousandth of an inch or so below. Check with a feeler gauge between the ruler and the needle or lever, and if the part is substantially above or below, something probably is very wrong, such as the number of the parts you were sold.

On float carburetors, check the float level carefully (7–39). Invert the float, which should be absolutely parallel to the carburetor surface. Most manufacturers provide a specification for float level, and you may be able to obtain it from the parts supplier. A common specification procedure is checking to see if a certain size drill will fit between the float and the carburetor surface. In the absence of specifications, hold the cover or bowl so that the float arm closes the needle valve (7–39). Then measure with a machinist's ruler from the top of the float to the bowl or cover surface at several points around the float. All measurements should be the same. If not, adjust the float needle tab.

On Briggs automatic choke systems, be sure to hook the choke spring onto the diaphragm as you reassemble.

CARBURETOR ADJUSTMENTS

Adjusting the carburetor is saved for last because this truly is the last job you should do. See 7–52 and 53.

Most owner's manuals provide the proper adjustment procedure, but if you've lost your little guide, here are some general rules that should help you.

1. Locate the idle-speed screw, which should bear against the throttle shaft or the linkage connected to it. To find the throttle shaft, look at the carburetor and see what linkage moves when you pull the trigger or throttle lever (7–13).

2. Next find the mixture screws. You'll probably find two, one alongside the other. See 7–52 and 53. A mixture screw has a slot head and a spring on its shank. On float carburetors, you may find one in the bottom of the fuel bowl (7–40, 41). If you find a second screw with a spring over the shank in another part of the bottom of the bowl, the odds are it's not a mixture

REPAIR PARTS KIT

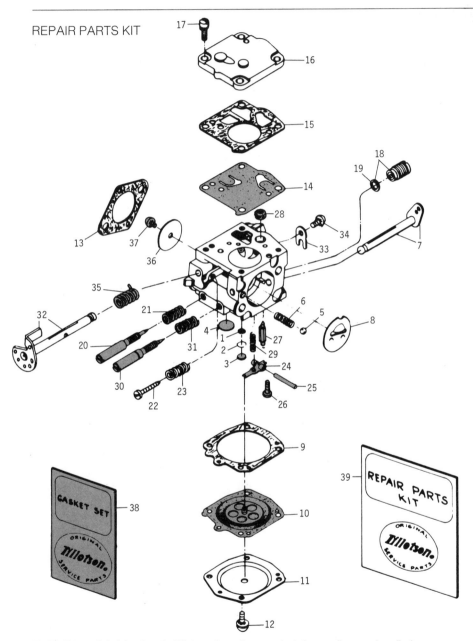

7-51. This exploded drawing of a Tillotson shows the many parts in a small gas engine, diaphragm carburetor. Parts that carry an asterisk (*) in the list on page 251 are included in the repair kit. As you can see, the repair kit includes welch plugs, diaphragms, mixture screws and a fuel needle. All gaskets are in a separate gasket kit. The two kits have all you should need to rebuild a carburetor at a fraction of the price of a new unit.

CARBURETOR PARTS SHOWN IN 7–51

KEY
NO.   DESCRIPTION

1.   Body Channel Screen
2.   Body Channel Screen Ret. Ring
3.   *Body Channel Welch Plug (Small)
4.   *Body Channel Welch Plug (Large)
5.   Choke Friction Ball
6.   Choke Friction Spring
7.   Choke Shaft & Lever
8.   Choke Shutter
9.   Diaphragm Gasket
10.   *Diaphragm
11.   Diaphragm Cover
12.   Diaphragm Cover Screw &
      Lockwasher (4)
13.   Flange Gasket
14.   *Fuel Pump Diaphragm
15.   Fuel Pump Gasket
16.   Fuel Pump Cover
17.   Fuel Pump Cover Ret. Screw &
      Lockwasher (4)
18.   Governor Valve Assembly
19.   Governor Valve Assembly Gasket

20.   *Idle Mixture Screw
21.   *Idle Mixture Screw Spring
22.   Idle Speed Screw
23.   Idle Speed Screw Spring
24.   *Inlet Control Lever
25.   *Inlet Pinion Pin
26.   *Inlet Pinion Pin Ret. Screw
27.   *Inlet Needle
28.   *Inlet Screen
29.   *Inlet Tension Spring
30.   *High Speed Mixture Screw
31.   *High Speed Mixture Screw Spring
32.   Throttle Shaft & Lever
33.   Throttle Shaft Clip
34.   Throttle Shaft Clip Ret. Screw
35.   *Throttle Shaft Return Spring
36.   Throttle Shutter
37.   *Throttle Shutter Screw & Lockwasher
38.   *Gasket Set
39.   Repair Parts Kit

(*)   Indicates contents of Repair Parts
      Kit. In the drawing, these parts
      are shown in gray tone.

screw, but a fuel bowl drain. If there's one at the center and a second at the side or at an obviously low point in the bowl, it's the drain; you can remove it to be sure. To identify each of the mixture screws, compare the thicknesses of the shanks. The thinnest is low-speed, the thickest is high-speed. With a screwdriver, turn each mixture screw in (clockwise) until it just seats; don't tighten. Then you are ready to proceed.

3. Turn each mixture screw head counterclockwise one turn. Turn the throttle speed screw counterclockwise until it is just off the point it touches on the linkage. Turn it clockwise until it just touches, then one turn more. These settings should get the engine to run, assuming nothing else is wrong, but they are not final.

4. Close the choke. Then start the engine, and try to keep it running at part throttle for at least three minutes to warm it up. If it stalls, turn the throttle speed screw clockwise a quarter of a turn and try again. On a chain saw, the chain may move very slowly while the choke is on; but once the engine is warm and the choke is off, it should not. Turn the choke off to see if the engine will idle; if not, turn the throttle speed screw a quarter to a half turn clockwise and try again.

5. With the engine idling, turn the low-speed mixture screw counterclockwise or clockwise to get the fastest idle. You may have to back off (turn counterclockwise) on the throttle speed screw to reduce idle speed to a normal level. In the absence of a tachometer to precisely measure engine speed, you'll have to judge by the sound of the engine. Although some

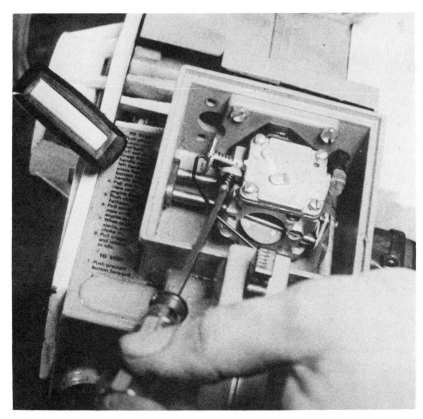

7–52. Turning the idle speed screw and the mixture screws on the left side just under it is the final step. Make small changes in screw positions and see how the engine responds. Large swings in position are not necessary.

7–53. This is a view of mixture screws from the housing exterior.

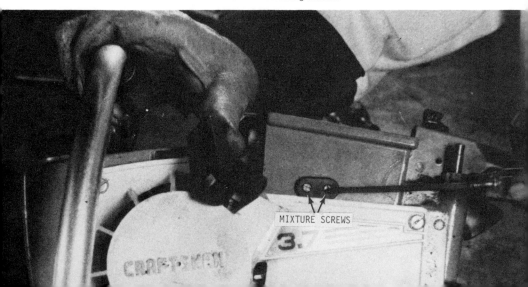

small gas engines will idle at as low as 200 rpm, the object isn't to set the idle at as low a speed at which the engine will sustain itself. Small-engine manufacturers specify from a low of about 600 to a high of about 2300; and they have good reasons, including the need to supply adequate engine lubrication. On a chain saw, set the speed for the point at which the chain just starts to move or moves very slowly. Then back off the throttle speed screw until it stops.

6. Run the engine at high speed. If you have a tachometer, work at 3000 rpm. Turn the high-speed needle clockwise until the engine just starts to falter, then counterclockwise until it runs smoothly, then the tiniest bit past that, which normally will be the best setting for the engine under load.

Always keep in mind these three general rules about carburetor mixture screw adjustments: 1) Turn the screw very slowly, to give the engine a chance to react to what you're doing. 2) Make very small changes. A $^1/_{16}$ of a turn is a significant amount of movement. 3) If you've turned a mixture screw three turns from the lightly-seated position, you've probably turned it much too far. On a McCulloch chain saw, for example, the best low and high-speed jet needle adjustments are between ½ and ¾ turn out. On a Sears mower with Tecumseh engine and Walbro carburetor, the best low-speed screw setting is usually 1¼ turns out; high speed, 1½ turns out.

CARBURETOR REINSTALLATION

Reinstalling the carburetor is generally very straightforward stuff, provided you have made some clear notes or committed the linkage arrangement to memory and know into which holes the governor and throttle springs go. See 7–54 through 59.

One important point: Do you remember when you put a straight edge against the carburetor surfaces (7–50) and tried to slip a feeler gauge underneath to be sure they were flat? Do the same to the engine or the gas tank surface to which the carburetor is bolted. If you find that the .002-inch gauge fits under, you're almost in a no-win situation, because this indicates that the mating surface is warped. You probably wouldn't spend the money to replace it—particularly if it's the engine.

Even on a Briggs setup, where the mating surface is the not-overly-expensive gas tank, you probably would find the thought of replacement more than an irritant. This is surely a case of try-the-second-gasket or nonhardening sealer, but a cautionary note is in order. You may be tempted to use gasket shellac to get a positive seal, but don't. Try two extra gaskets and nonhardening sealer if you must. If you use gasket shellac you can be sure that Murphy's Law will require that for some reason you'll have to take the carburetor off very soon, and you won't be able to, without ruining something.

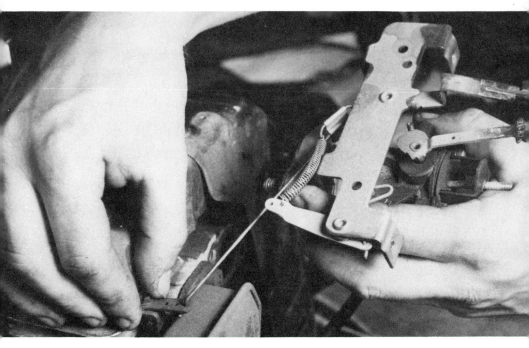

7-54. Reinstalling a carburetor shouldn't be tricky if you've carefully noted what goes where. For example, there's a choice of three holes for the thin rod, which (on this unit) goes into the end hole.

7-55. The carburetor is held by Phillips-head screws. To tighten them, mount a pliers on a screwdriver shank.

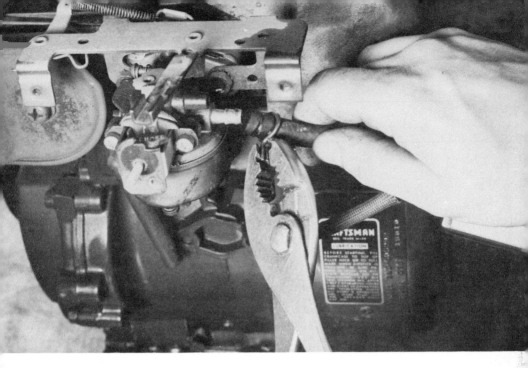

7–56. Remember to fit the fuel line and reposition the spring clamp.

7–57. Ready to install the cover? Don't forget the wire to the magneto kill switch.

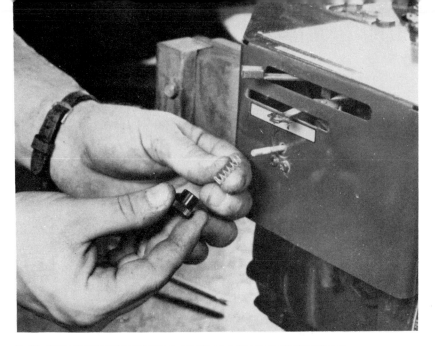

7–58. Before threading the primer knob onto the projecting shaft, slip the spring on.

7–59. The knob's back on and the job is done.

7-60. SERVICE TIPS FOR DIAPHRAGM CARBURETORS

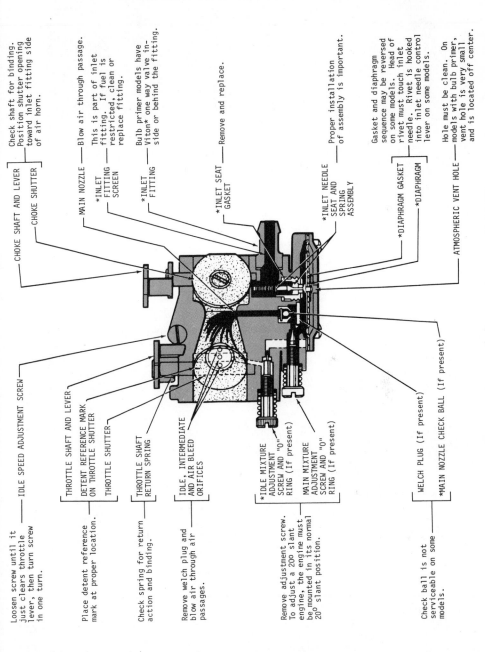

Check shaft for binding. Position shutter opening toward inlet fitting side of air horn.

CHOKE SHAFT AND LEVER

CHOKE SHUTTER

MAIN NOZZLE — Blow air through passage.

*INLET FITTING SCREEN — This is part of inlet fitting. If fuel is restricted, clean or replace fitting.

INLET FITTING — Bulb primer models have Viton one way valve inside or behind the fitting.

*INLET SEAT GASKET — Remove and replace.

*INLET NEEDLE SEAT AND SPRING ASSEMBLY — Proper installation of assembly is important.

*DIAPHRAGM GASKET — Gasket and diaphragm sequence may be reversed on some models. Head of rivet must touch inlet needle. Rivet is hooked into inlet needle control lever on some models.

*DIAPHRAGM

ATMOSPHERIC VENT HOLE — Hole must be clean. On models with bulb primer, vent hole is very small and is located off center.

Loosen screw until it just clears throttle lever, then turn screw in one turn.

IDLE SPEED ADJUSTMENT SCREW

THROTTLE SHAFT AND LEVER

Place detent reference mark at proper location.

DETENT REFERENCE MARK ON THROTTLE SHUTTER

THROTTLE SHUTTER

THROTTLE SHAFT RETURN SPRING

Check spring for return action and binding.

IDLE, INTERMEDIATE AND AIR BLEED ORIFICES

Remove welch plug and blow air through air passages.

*IDLE MIXTURE ADJUSTMENT SCREW AND "O" RING (If present)

MAIN MIXTURE ADJUSTMENT SCREW AND "O" RING (If present)

Remove adjustment screw. To adjust a 20° slant engine, the engine must be mounted in its normal 20° slant position.

WELCH PLUG (If present)

*MAIN NOZZLE CHECK BALL (If present)

Check ball is not serviceable on some models.

*NONMETALLIC ITEMS: CAN BE DAMAGED BY HARSH CARBURETOR CLEANERS

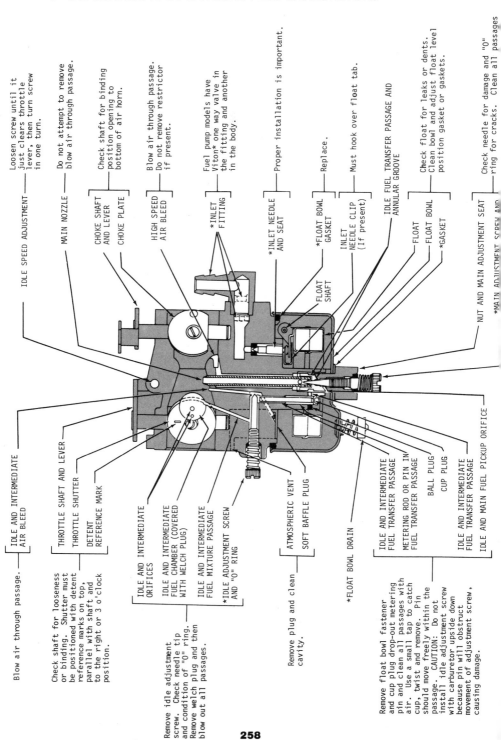

Blow air through passage.

Loosen screw until it just clears throttle lever, then turn screw in one turn.

Do not attempt to remove blow air through passage.

Check shaft for binding position opening to bottom of air horn.

Blow air through passage. Do not remove restrictor if present.

Fuel pump models have Viton* one way valve in the fitting and another in the body.

Proper installation is important.

Replace.

Must hook over float tab.

Check float for leaks or dents. Clean bowl and adjust float level position gasket or gaskets.

Check needle for damage and "O" ring for cracks. Clean all passages

IDLE SPEED ADJUSTMENT

MAIN NOZZLE

CHOKE SHAFT AND LEVER

CHOKE PLATE

HIGH SPEED AIR BLEED

*INLET FITTING

*INLET NEEDLE AND SEAT

*FLOAT BOWL GASKET

INLET NEEDLE CLIP (If present)

IDLE FUEL TRANSFER PASSAGE AND ANNULAR GROOVE

FLOAT

FLOAT BOWL

*GASKET

FLOAT SHAFT

NUT AND MAIN ADJUSTMENT SEAT

*MAIN ADJUSTMENT SCREW AND

IDLE AND INTERMEDIATE AIR BLEED

THROTTLE SHAFT AND LEVER

THROTTLE SHUTTER

DETENT REFERENCE MARK

IDLE AND INTERMEDIATE ORIFICES

IDLE AND INTERMEDIATE FUEL CHAMBER (COVERED WITH WELCH PLUG)

IDLE AND INTERMEDIATE FUEL MIXTURE PASSAGE

*IDLE ADJUSTMENT SCREW AND "O" RING

ATMOSPHERIC VENT

SOFT BAFFLE PLUG

*FLOAT BOWL DRAIN

IDLE AND INTERMEDIATE FUEL TRANSFER PASSAGE

METERING ROD OR PIN IN FUEL TRANSFER PASSAGE

BALL PLUG

CUP PLUG

IDLE AND INTERMEDIATE FUEL TRANSFER PASSAGE

IDLE AND MAIN FUEL PICKUP ORIFICE

Check shaft for looseness or binding. Shutter must be positioned with detent reference marks on top, parallel with shaft and to the right or 3 o'clock position.

Remove idle adjustment screw. Check needle tip and condition of "O" ring. Remove welch plug and then blow out all passages.

Remove plug and clean cavity.

Remove float bowl fastener and cup plug drop-out metering pin and clean all passages with air. Use a small tap to catch cup, twist and remove. Pin should move freely within the passage. CAUTION: Do not install idle adjustment screw with carburetor upside down because pin will obstruct movement of adjustment screw, causing damage.

Mastering Lubrication and Maintenance

If you really want your small gas equipment to last, you can give it a lot of tender loving care. I have seen recommended maintenance charts that would have you spending more time taking care of the equipment than using it. Yes, the equipment might last twice as long, but unless you're retired and looking for ways to spend your time, the super-maintenance schedule just isn't practical. You're a homeowner with a lot of things to do around the house. You want to attend to your small gas engine equipment because you need it at home, and not lingering in some repair shop. And this equipment may be the third largest investment you've made—right after your home and your car.

A no-maintenance policy is a sure way to send a $150 to $600 mower to the scrap heap in two years. Depending on the use they get, the $150 to $250 chain saw and the $200 to $600 snow blower might not go that much longer. When you're talking about an investment that may approach or exceed four figures, that's not small potatoes, and small gas engine repair shops in suburban areas get at least $12 an hour for their time.

Let's look at an adequate maintenance schedule, therefore, the kind that will keep the small gas engine going—without keeping you going.

THE OVERALL ENGINE PROGRAM

The maintenance program for a small gas engine depends on the type of use it gets. The mower gets regular use during the season, the blower frequent but somewhat irregular use. The chain saw gets a real workout when you're on a project, but it also can sit for a long period. The mower, therefore, should get major maintenance at the start of the season and once again somewhat past the half-way mark.

The blower should get major maintenance at the start of the season, then once more after about ten outings. If the snow season is light, you may be able to get through an entire season without further service. The best thing you can do for the blower is to store it under a plain blanket in a closed, unheated garage.

The chain saw should get some maintenance just before a period of heavy use (such as cutting firewood for winter, building a fence) and major maintenance at least once a year.

Cleaning Away Debris

You could remove covers and really give the engine a thorough cleaning after every use. But aside from the chain saw, which normally gets heavy but irregular use, this isn't really necessary. The mower is the second major collector of debris. But a quick, light cleaning after each use and a major cleaning every fourth use (roughly once a month during the summer) will prevent it from getting severely clogged mowing the typical suburban quarter to half-acre lot. See 8–1 through 8–3. The snow blower accumulates virtually no debris on a tar or concrete driveway; but on a gravel or unpaved driveway or on a walking path, it may pick up a fair amount. In this latter case it's a good idea to check to see that there is no hard debris in the area of the auger after each use.

Note: Some lawn mowers have a so-called cleaning port, to which you attach a water hose and then let the combination of water power and a spinning engine clean the debris away. These port setups do only a fair job of cleaning at best and they can let water seep into spots it shouldn't, and cause rusting. If you want to get the benefit of this port arrangement, you should have a compressed air tank to be able to dry off the machine. A tank costs about $20 to $30 and can be refilled at a service station pump. It's an extremely handy item for cleaning and drying all kinds of parts, and for inflating tires.

Air Filter

The lawn mower's carburetor air filter should be cleaned or replaced when you perform routine major maintenance at the beginning and past the halfway mark of each season. This is described in Chapter 7. Also you should visually inspect and clean the filter if necessary when you clean the chassis.

The chain saw carburetor air filter should be serviced as part of its annual maintenance and inspected—and serviced if necessary—just before a period of heavy use.

Keep areas
within
heavy line
clear
of all
debris

CLEANING AWAY DEBRIS

8–1. On a vertical crankshaft engine such as on a rotary lawn mower, areas within heavy lines should be cleared carefully of all debris.

Clean out
chaff and
dirt

8–2. The arrows here point to sections of a horizontal crankshaft engine that are likely to accumulate debris.

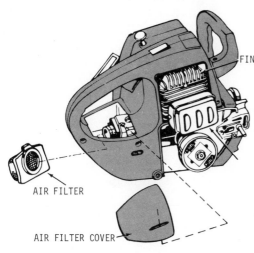

FINS

8–3. On a chain saw, periodically remove the housing covers, clean the fins, and service the air filter.

AIR FILTER

AIR FILTER COVER

The snow blower air filter should be serviced as part of pre-winter maintenance.

Ignition System

The ignition system should be inspected and plugs, points and condenser replaced at the start of the season in the case of the mower and blower, and as part of annual maintenance on the chain saw. In addition, the spark plug should be inspected, cleaned with a wire brush, and the gap adjusted to specifications during the later-in-the-season maintenance on a mower and blower. On the chain saw, the plug should be checked prior to any period of heavy use, because the two-cycle engine is more likely to deposit oil on it.

If you mow a lawn much over a half-acre or blow a long driveway with heavy snow, so that the machine is operating many hours, clean the plug and check the breaker points after 50 hours of use. Adjust or replace them as necessary.

Carburetor

As part of preseason and later-in-the-season maintenance on a mower or blower, spray the linkage at the carburetor with a penetrating solvent. On a chain saw, spray as part of annual maintenance and before a period of heavy use. Do not oil the linkage, for oil provides a sticky surface to which dirt can adhere, causing the linkage to bind. If the throttle or choke on a mower or blower is controlled by cables, disconnect them; pull them out from their housings and coat them with automotive speedometer cable lubricant, which is graphite-based.

Engine Mounting Bolts

The bolts that hold the engine to the chassis on a mower or blower should be checked for tightness during preseason and later-in-the-season maintenance.

Carbon and Valve Job

The carbon and valve job, described in Chapter 5, is not a routine maintenance item. But almost any four-cycle small gas engine develops tired blood and often benefits from this work every other year as part of preseason maintenance. It will help restore some of the power that is lost as an engine wears. If you really give your engine a workout, perhaps 100 hours a season, you should consider the job.

Preparation for Seasonal Storage

Before storing a small gas appliance for the season, drain the fuel tank completely, then run the engine until it has burned off all the gasoline. If you fail to do this the fuel will interact with moisture and form a gummy substance that will cause clogging in the fuel lines and carburetor, and cause moving parts to stick. Also remove the spark plug and pour four or five tablespoons of oil into the cylinder (through the plug hole) to minimize rust and corrosion. Reinstall the plug, but leave the wire disconnected. Crank the engine with the starter three or four times to circulate the oil. Note: Next season the engine may blow some blue smoke during the first few minutes of operation, as it burns the oil.

Lightly oil the wheels at the shafts on which they're mounted and all other parts that move, such as chains and blower auger mountings. To retard rusting, put a few drops of light household oil on any external moving part.

Oil Change and Gasoline

Gasoline and oil are perhaps the most controversial aspects of operation and maintenance of gasoline engines, regardless of size. Let's begin with gasoline.

The small gas engine manufacturers do not disapprove of use of unleaded gasoline, but until 1974 it was virtually unavailable except as a premium-priced product in Amoco and a few small-brand stations, plus marine fuel outlets. As a result, virtually no one used it in small gas engines.

Now that unleaded regular gasoline is so widely available for automobiles, it will doubtlessly be coming into greater use in small gas engines around the home. In fact, many people see that it will extend the life of the spark plug and the engine oil by reducing lead contamination and eliminating the lead deposits that create the need for carbon and valve jobs.

Before you make the switch, however, you should consider the fact that lead in gasoline provides a special form of lubrication for the valves and seats in a four-cycle engine. If you use lead-free gasoline on an occasional basis, even up to half the time, there should be no problem. However, field-test results past that point are lacking at this time, so proceed cautiously.

There is no octane rating problem with unleaded gasoline, for even the lowest (86 to 87 combination rating, equal to about 91 on the best known research rating scale) is substantially more than the typical small gas engine needs.

On a two-cycle chain saw or lawn mower, the reed valves in the crankcase do not need lead lubrication, so you can safely switch if you wish.

Filling the Tank

It might seem that filling the gas tank is such an easy thing that it's not worth talking about. Actually, the way you put gasoline in the tank is important, and the best procedure depends on usage and storage.

An important rule is never transfer gasoline from mower to blower or vice versa. Gasoline is formulated for specific seasons, and the gasoline in winter is made with a lot of what are called "light hydrocarbons," which means that the fuel vaporizes easily for combustion in cold weather. Summer gasoline is made with more "heavy hydrocarbons," because summer heat will vaporize them easily. If you use summer gas in a snow blower, you'll have a devil of a time trying to start it in cold weather. If you use winter gas in a mower, most of the fuel will evaporate before it reaches the combustion chamber; and although the engine will start, it won't develop much power. In each season, of course, you can pour the same fuel into the two-cycle chain saw that you are using in the other appliance, after mixing the fuel with the proper proportion of oil for the saw. Never transfer fuel from a chain saw to one of the others, for they are not designed to run very well on a mixture of gas and oil, and even if they do run, the oil will leave unacceptable deposits.

If you store your lawn mower indoors during the summer, it's not a good idea to keep the tank full. If indoors, the heat will evaporate enough fuel to create odors and possibly a fire hazard. Outdoors this hazard does not exist for all practical purposes, but keep the mower in a shady spot so that sunlight doesn't cause noticeable fuel losses due to evaporation.

The snow blower should be kept indoors and covered with a blanket during cold weather for easier starting, and the best place is an unheated garage. If this sounds strange to the man who has a heated garage available, here is the reasoning. If the garage is warm, the "light hydrocarbons" will easily evaporate, causing both loss of fuel and poor performance, for these are the parts of the gasoline that contribute most to cold weather starting and operation. Of course, you will start the engine as it comes immediately from the warm garage, but it may run poorly until it warms up fully and can use the heat it produces to vaporize the fuel. In a heated garage, you also could get fuel odors and possibly a fire hazard.

The unheated area is best because it will still be 20 to 30 degrees warmer than the outdoors, and the blanket adds some insulation over the engine. Because the temperature is not as high as in the house, there will be no significant fuel evaporation. Keep the tank full to minimize condensation of moisture.

Inasmuch as the overwhelming majority of Americans have unheated garages, the discussion above is largely academic. If you have a heated garage, you should turn down garage heat.

Because the chain saw gets much irregular use, fill the tank only with as much gasoline-oil mixture as you need. To prevent gum formation, drain the tank whenever the saw is to be put away for more than a few weeks.

The correct oil for two-cycle engines is not the correct oil for four-strokers. The two-stroke engine burns oil with gasoline, and its higher oil consumption would create a combustion chamber deposit problem if the oil recommended for four-strokers were used. The four-stroker, by contrast, must provide lubricant for a camshaft and valve parts that are not easy to oil. Therefore the oil must contain additives that provide adequate lubrication in small quantities. Valve system parts also are somewhat sensitive to sticking from sludge that forms in the oil, so the oil must be formulated with additives that keep the sludge from building up and depositing itself on the valves.

Four-Cycle Oil

The four-cycle engine manufacturers all recommend an oil specified "for Service MS," or "SC" or "SE," which are automotive oils. MS is a rating of certain qualities in the oil, qualities achieved by the addition of certain additives. At one time MS was the highest rating, but the addition of emission controls on cars made the oil industry change the additive formulations, and a few years ago, it changed its lettering system. MS is now SC, although both sets of letters usually are found on the can. SC is approved for use in cars up to 1967. SD is for 1968–70, and SE is for 1971 cars and on.

The reason for this explanation is that you might be able to save some money if you buy your oil right. The typical lawn mower or snow blower engine really doesn't need what's recommended for a late-model car. The operating stress placed on the small gas engine is far less than anything a car engine must undergo, even a 1967 model. Therefore, an MS-SC of just plain MS or SC is really all your small gas engine needs, and to spend money for an oil that meets SC, SD, and SE standards is a waste unless you buy your engine oil by the case and want an oil that will work in your car and in your mower and blower.

If you change the oil only in your small gas machines and leave the car to the service station, buy the least expensive type. The only caution is to stick with name brands, for there is no industry enforcement of these lettering guides, so the lettering is only as meaningful as the refiner's reputation.

Letters refer to the general service for which the oil is recommended. You still have to pick oil of the proper thickness for good lubrication. If you use too thick an oil in winter, it won't flow freely enough to lubricate the engine, particularly the valve system. If you use too thin an oil in summer, the heat will thin the oil out so much that it will flow too freely, and parts will not be adequately lubricated.

Oil thickness is also called "viscosity," and you must pick oil of the correct viscosity for the climate. The thickness numbers range from five up to 40 or even 50 and 60 and carry the prefix "SAE" which stands for Society of Automotive Engineers, the organization which sets the thickness specifications for each number. If the letter "W" follows a number, it means that the viscosity was measured with the oil cold. A 10W-40 oil, therefore, has a 10 thickness when cold, a 40 thickness when hot. The oil doesn't get thicker when hot; it just thins out a lot less than a normal 10 oil. There is a permissible range, and a choice of oils. Here it is:

For temperatures consistently over 32 degrees (lawn mower operation), use SAE 30 (the single grade is the least expensive), SAE 10W (the next least expensive) or SAE 10W-40 (the most expensive).

For temperatures consistently below 32 degrees but not regularly below zero, use SAE 10W (the cheapest), SAE 5W-20 (next up in the price scale) or 5W-30 (the most expensive).

For temperatures consistently below zero, use 10W diluted with three ounces of kerosene per quart. Briggs and Stratton permits the use of 10W-30 diluted with kerosene if the single weight 10W is unavailable; Tecumseh does not.

The oil should be drained when the engine is hot, which means after the engine has run for at least 15 minutes (see 8–4 and 8–5). Hot oil is thinner; therefore, more of it will flow out when the drain plug is pulled. In most rotary mowers, the drain plug is located close to the blade, so you might consider oil change time a suitable occasion to take off the blade and get it

---

DRAINING OIL

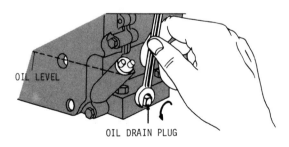

OIL LEVEL

OIL DRAIN PLUG

8–4. Oil should be drained when the engine is hot. If there is no drain plug, twist off the fill plug or cap. Then pick up the engine and invert it.

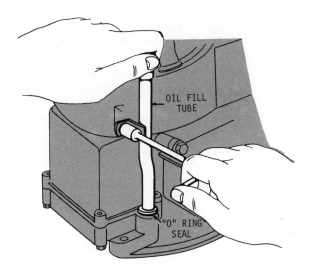

8–5. On Briggs engines with the extended oil fill tube and dipstick, always check to make sure there's no leak at the O-ring seal. A leak there (or at the upper end of the dipstick tube, if the dipstick is loose) can cause the engine to smoke.

sharpened. On mowers and blowers, change the oil as part of the preseason maintenance and during the later-in-the-season maintenance.

Two-Cycle Oil

There are many brands of oil recommended for two-cycle engines, generally under the name "outboard motor oil." Virtually all chain saw manufacturers except McCulloch prescribe SAE 30 nondetergent outboard motor oil for service "MM" or "SB" (equivalent designations) in a ratio of 16 parts of gasoline to one of oil (a half pint of oil per gallon), which must be thoroughly mixed prior to pouring into the tank. If it's near freezing outside, keep the oil warm indoors. Mix all the oil with half the gasoline in a container (shake vigorously for several minutes; don't just slosh it around). Add the remainder of the gasoline and again shake vigorously for several minutes.

McCulloch recommends a nondetergent SAE 40 in a 16–1 ratio or its own brand of two-cycle in a 40–1 ratio (3½ ounces per gallon), claiming that its own oil is designed for lightweight two-cycle engines, hence can provide the lubrication the engine needs with half the oil. Less oil means cleaner burning of the gas-oil mixture. If you find a reputable refinery brand of 40-1 oil marketed for use in chain saws, it should do the job too. Regardless of oil, McCulloch also recommends the vigorous mixing procedure.

Properly mixed fuel stays mixed even in freezing weather. If you don't mix oil and gas thoroughly, there will be large enough oil droplets to cause significant carbon deposits in the combustion chamber. These deposits get very hot during combustion and can cause ignition of the air-fuel mixture at the wrong time.

If you have a larger chain saw with a transmission in a housing filled with oil, use the specific gear oil recommended by the chain saw manufacturer, and that generally is not the same as engine oil. McCulloch, for example, recommends SAE 140 gear oil for some models, automatic transmission oils for others. Other manufacturers' recommendations vary widely too, with some calling for SAE 20 engine oil in cold weather, 30 in warm; SAE 90 gear oil in cold weather, SAE 140 gear oil in warm. Additionally, some manufacturers specify a general purpose gear oil; others insist on a gear oil with "extreme pressure" qualities ("EP" is the industry abbreviation), which means the oil has special additives for special lubrication situations.

Transmission oil should be changed as part of annual maintenance. Follow the manufacturer's specified procedure for filling the transmission. Do not operate on the assumption that oil is good and you can t put in too much of a good thing. An overfilled transmission does not have enough room for expansion of the oil from the heat of operation, and if the oil gets hot enough, it will expand sufficiently to force its way out past the seals, damaging them in the process.

Note: Some gear-driven saws do not have an oil-filled housing. Instead, exposed gears are lubricated with a few drops of SAE 30 nondetergent engine oil before each use.

The reason for the differing recommendations on type of oil is the metallurgy of the transmission and the loads imposed on the gears. The wrong gear oil will definitely cause the premature demise of the transmission.

DRIVEN MECHANISMS

There is more to a small gas machine than the engine and the transmission. Proper lubrication and maintenance starts with the engine but should not end there. You can easily ruin the engine of a rotary mower by cutting with a blade so nicked that it is out of balance. By trying to cut wood with a dull or inadequately adjusted or lubricated chain, all you're doing is overstraining a small engine, which will respond to this treatment by wearing out a lot sooner.

In addition to the major engine maintenance prescribed earlier in this chapter, you should give the driven mechanisms and accessories an equal amount of attention and care.

It would be impossible to discuss all the variations on the driven mechanisms and the accessories used on mowers, blowers and chain saws. Your owner's manual is your best guide to what the manufacturer of your equipment has built and what, in his judgement, should be cleaned, lubricated and adjusted. This section is merely intended to call your attention to some things that may not appear in the manual and to expand on a few things that the manufacturer may have given short shrift.

CHAIN SAW

The heart of the chain saw may be its two-cycle engine, but the chain and its associated parts are surely the muscle, and they require periodic inspection, lubrication and adjustment.

The lubrication is done by the oiling mechanism, which pumps oil from the reservoir. You may use one of the chain oils sold by the saw manufacturers or a nondetergent SAE 30 recommended for service or SB, which is the same oil recommended by most saw manufacturers for the gasoline mix. The advantage of the SAE 30 is that on most saws a single oil is all you need. Oil in the chain reservoir does not deteriorate measurably, so you never have to drain the reservoir. You should, however, always keep it topped up to minimize moisture formation. Note: In below-freezing weather, dilute the oil with kerosene, in equal portions, or switch to SAE 10W, a much lighter oil, to insure adequate flow to the chain.

Some saws have automatic and manual oiling systems, with both feeding from the same reservoir. The automatic oiler is adjusted for average cutting conditions, and although there is an adjustment screw to permit you to increase the flow, it should not be touched simply because you've got some heavy-duty cutting to do. Careless adjustment could increase the automatic flow rate to automatically empty a full reservoir in a matter of minutes. If you have some heavy cutting, you're better off hitting the manual oiler button a few times during the job.

You should check the manual oiler before starting the saw. Just pump the button and look for oil at the discharge point, which typically is between the chain and bar, at the top of the bar just forward of the centrifugal clutch sprocket cover. Clean the discharge hole with a thin piece of wire if necessary.

Chain Saw's Clutch Sprocket Bearing

The centrifugal clutch that joins the engine to the chain in a chain saw needs no maintenance, but the bearing on which the chain sprocket turns should be cleaned in solvent and packed with a high-temperature light grease, such as Lubri-Plate.

Adjusting Chain Tension

Also check chain tension. When you hold the bar nose up, the chain should appear snug, even though you can still pull it easily around the bar. See 8–6. Always check tension before using the saw, or after it has cooled. The adjustment is simple enough. If necessary slacken the chain bar nut or nuts. Then turn the adjusting screw, as shown in 8–7. This draws the bar farther out along its elongated slots.

8–6 and 8–7. The chain should have enough play for you to see most of the base of each cutting link when you pull outward, below left, and you should be able to draw it freely around the guide bar. To adjust the chain, slacken the chain bar nut(s), then turn the adjusting screw as shown. Turn the screw clockwise to increase tension, counterclockwise to decrease it.

Breaking In a New Chain

Installation of a new chain is a somewhat evident procedure if you have studied Chapter 5. To "break in" the chain you should run it at slow speed for several minutes, making sure oil is being delivered to it. If you haven't got an automatic oiler, hit the oiler button a couple of times before starting, then a couple of times per minute. Open the throttle by pulling the trigger to the halfway point and run the chain for another few minutes, making sure you hit the oiler button two to three times per minute. Stop the engine, adjust chain tension if necessary, then use the saw for 10 to 15 minutes. Again stop the engine and check tension. Repeat tension checks every hour for the first three to four hours. Note: Chains aren't cheap. Have dull ones professionally sharpened to save yourself some money.

Automatic Sharpeners

A built-in chain sharpener is really a misnomer, for the device is nothing more than a grinding bar that removes burrs and nicks. Sharpening is another operation that should be done at least once a year, but because no one does the same kind of cutting the same number of hours as anyone

---

8–8 and 8–9. The snow blower gearcase in the left-hand photo is partly covered by a olate, but it is somewhat exposed to the rigors of the elements. The screwdriver points to the grease fitting on the shaft of the blower. All areas subject to corrosion should be packed with grease.

else, the only other rule is to sharpen the chain when it's dull. Don't just try to push harder on the saw.

The chain saw manufacturers sell the special tools and instructions necessary to sharpen blades on their particular saws and adjust what are called "depth gauges," but these are jobs best left to a professional.

MOWER AND BLOWER LUBRICATION AND CLEANING

Thorough cleaning of the mower or blower chassis is a basic but essential procedure. You can save yourself some of the tedium of digging out dirt by blowing it away with a compressed air tank.

The mower chassis should be given a thorough cleaning every fourth use under normal conditions. At the start of the mowing season, during second-half maintenance and before storing the machine away for the winter, you should also lubricate all moving chassis parts, particularly the wheels, the pulley shafts, or chains—in fact, the joint of every part that moves. Use a household penetrating oil.

The blower is less likely to need the chassis cleaning that the mower does, but it certainly needs the same kind of lubrication to prevent parts from rusting as they get waterlogged moving through the snow.

If you have a blower with an auger gear set that is covered but not in a closed housing, the parts may require greasing, as shown in 8–8 and 8–9. There are so many possible designs that the best procedure is to check the owner's manual.

Reel Mower Cutters

The reel mower should be checked for proper adjustment of the reels and cutting bar at both major maintenance sessions. The check is simple enough: Slip a piece of typing paper between the reels and bar at each side and in the center; turn the reel manually, as in 8–10, and it should cut the paper neatly. If it doesn't, and the reels and bar are sharp, you can adjust the reels, as in 8–11, to bring them closer to the bar if necessary. Don't bring them into tight physical contact or you'll wear off the fine cutting edges. If after adjustment the center of the cutting bar is too close or too far away, some mowers have a center spring-loaded screw to force the bar down toward the reels, as shown in 8–12. Leave blade sharpening to the professional.

Also check chain or belt tension at the same time. As shown in 8–13, the chain or belt should deflect no more than about half an inch between pulleys or sprockets. The adjustment may be as simple as slackening the mounting bolts of the engine and moving the engine back (to increase tension) or forward (to reduce it). Or there may be an idler pulley on a spring or on a bolt in an elongated hole, as in 8–14.

8–10 and 8–11. Check a reel mower's cutting ability with a paper inserted at the ends of the cutting bar, as shown above left. To bring the reel on this model closer to the cutting bar, slacken the small nut and turn the large one. Make the adjustment on each side. (Some reel mowers have shim adjustments or one of several other adjusting arrangements.)

8–12 and 8–13. If the reel mower has an adjusting-screw and concave-spring arrangement, as shown below left, you can check the cutting ability of the blades in the center and make an easy adjustment if necessary. Turning the screw down forces the cutting bar toward the reel; turning it out allows the bar to move from the reel. You can check chain tension at any convenient opening; the chain should deflect about half an inch.

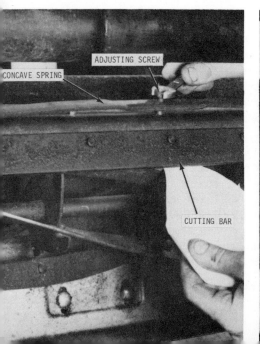
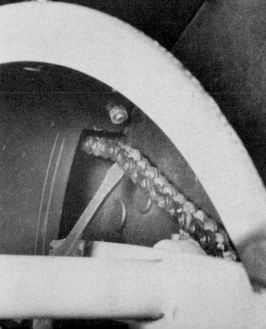

CONCAVE SPRING
ADJUSTING SCREW
CUTTING BAR

BRACKET BASE

BRACKET

ELONGATED HOLES

8–14. There are various adjustments for chain tension on mowers and blowers, including moving the engine along elongated holes. In this case, there's a bracket held by nuts and bolts through elongated holes. To increase chain tension, slacken the nuts and move the bracket rearward. To reduce tension, move it forward.

Rotary Mower Blade Slowdown

Some better rotary mowers have a blade slowdown feature that stops the blade within a few seconds after the self-propulsion clutch is disengaged. The usual system has a drive belt and an idler pulley that is pulled tight when the mower handle is lifted, dragging a pulley on which the blade is mounted to a stop. Or there may be a friction band wrapped around the belt at the blade pulley; and when you lift the handle you tighten the band against the belt, stopping the blade pulley and therefore the blade. In either case the belt should be checked periodically and replaced when it shows surface cracks. An elongated hole arrangement on the idler pulley mounting bolt is the usual method of slackening tension for belt removal or increasing it. The friction band normally is adjustable by tightening it down on a threaded rod. To check your adjustment, just make sure the blade is absolutely free to spin when the handle is in the self-propulsion position.

Appendix

DECIDING TO REPAIR OR REPLACE

Machines don't last forever, and unless you recognize the economics, you can waste money trying to refurbish equipment that is better replaced. It's also easy to mistakenly decide to junk the appliance and look for a replacement.

Knowing when to repair and when to replace is not an exact science, but some useful rules of thumb may help you. Here are guidelines for making the decision when the engine fails.

Rule Number 1 is to know what's wrong and how much the repairs will cost. The troubleshooting charts in Chapter 4 will help you figure out what's wrong. If you can handle the problem, a quick call to the parts supplier will tell you what the parts cost. Then, no matter how old the machine is, if all it needs is one of the repairs covered in this book, you can do the job yourself, for you will be saving the labor cost, generally the lion's share of the package.

When you need a major repair you cannot do on an older machine, the decision-making becomes a bit tricky.

Chain Saws

On most chain saws, for example, the cost of an engine overhaul is no more than two-thirds the price of a comparable new machine. (That's for the worst situation, in which the engine block must be replaced along with crankshaft, connecting rod and piston.) If you've been good about lubricating the saw's engine, there is a very good chance that it may need nothing more than new crankshaft bearings and piston rings, and perhaps a light grinding of the cylinder with a tool called a hone. This could be less than half the price of a new machine. If the saw is only a few years old and in otherwise good condition, getting professional repairs could give you an almost-new machine for far less than a new one. Even if the engine is virtually destroyed, two-thirds is less than three-thirds, and if the failure was a fluke on a nearly-new machine (but unfortunately out of warranty), an engine overhaul probably would pay. You may be able to save $15 or more by removing the engine from the chassis and reinstalling it yourself because this saves the shop an hour and change. Check with the shop first on this, however, for some shops charge a flat rate.

Mower and Blower Engines

You often have a choice with the four-cycle engines on mowers and blowers: repair or replace the engine. A factory-rebuilt and guaranteed engine is about 60 to 70 percent of the price of a new rotary mower and as little as 40 to 50 percent of the price of a new blower or reel mower. If the deck (chassis) and driven mechanisms are in good shape, the saving is worthwhile. This is particularly true of a rotary mower, for if you lubricate it and clean the chassis regularly, about all that can happen is that a tire will go flat or that the deck will crack. Inasmuch as these things could happen to a brand-new machine as well, a rebuilt engine is an excellent investment that could give you very close to what you'd have with a new appliance.

In the case of a reel mower, although the engine is important, there is a lot more to the machine. So if you need only a rebuilt engine, rebuilding pays, even if the reel is four or five years old. A well-maintained reel can last for a decade. A maintained rotary, by contrast, is lucky to survive four years, although its failure is normally limited to the inexpensive-to-replace engine.

When to Replace the Machine

You should know your machine, and if you have given it reasonable care, don't hesitate to invest in an engine overhaul or a rebuilt engine. If the appliance has been abused, however, don't hope an engine will cure all its ills. If you know you've banged around the chassis of a rotary mower, it's just a matter of time before the cocked chassis will bring on other failures. If the wheels turn a bit stiffly, they may soon need replacement. Wait for a sale, buy new equipment, and this time give it the care that will make a major repair, if ever needed, a sensible investment.

MANDATORY SAFETY CONTROLS

Late-model walk-behind rotary mowers include mandatory safety systems to help prevent injury from the spinning blade. If you release the operating handle (normally held against the frame handle), the ignition is killed and a spring-loaded linkage moves a brake shoe against the flywheel, stopping both engine and blade. Alternatively, some premium mowers use a brake *clutch* between blade and engine. Releasing the handle stops only the blade, allowing the engine to keep running so you don't have to restart it.

If the blade continues to turn after you release the handle on safety-equipped machines, check the spring-loaded cable and linkage. If they operate normally, replace the brake shoe as shown in 9-1 and 9-2. On mowers with a brake clutch, also check the linkage, cable, and spring. If these operate normally, but the blade turns when the handle is released, replace the brake-clutch assembly — available as a replacement kit — as shown in 9-3 through 9-5.

The brake-shoe linkage typically has a kill switch to stop the engine. If this switch fails, it may prevent the engine from starting. Disconnect the switch, and if the engine now starts, replace the switch. If the engine still won't start, check for spark as explained on page 45. If you get no spark, inspect the flywheel for cracks, a loose magnet, or a severely worn keyway (see 9-6). If the kill switch and flywheel are good, install a new electronic ignition module.

9-1. Typical linkage arrangement, which operates kill switch and pushes brake shoe against flywheel to stop engine, is shown outside mower.

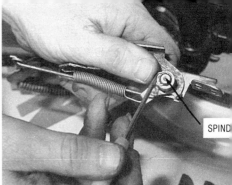

9-2. After removing snap ring, lift the brake shoe off its spindle for replacement.

9-3. Disconnect spark-plug wire, then begin brake-clutch replacement by loosening the bolt in the center of the blade. Do this with blade *on*, using blade to hold brake clutch in position. Then remove the blade.

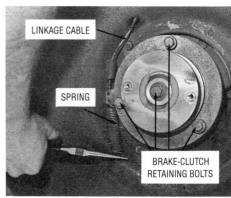

9-4. Disconnect spring and linkage cable, then remove the four mounting bolts.

9-5. Brake-clutch assembly comes off as shown. *Do not lubricate ball bearings as they are part of the clutch and, if lubed, will attract dirt and malfunction.* Pry out and replace the key, then install new parts.

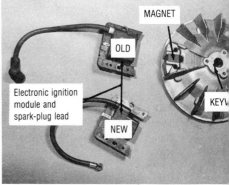

9-6. If you're not getting a spark, and the kill switch is good, check the flywheel and keyway for damage. Also check for a loose flywheel magnet. If these are good, install a new electronic-ignition assembly.

INDEX